Probability

Key Concepts in Philosophy

Guy Axtell, *Objectivity*
Heather Battaly, *Virtue*
Lisa Bortolotti, *Irrationality*
Ben Bradley, *Well-Being*
Joseph Keim Campbell, *Free Will*
Roy T. Cook, *Paradoxes*
Douglas Edwards, *Properties*
Ian Evans & Nicholas D. Smith, *Knowledge*
Bryan Frances, *Disagreement*
Amy Kind, *Persons and Personal Identity*
Douglas Kutach, *Causation*
Carolyn Price, *Emotion*
Daniel Speak, *The Problem of Evil*
Mathew Tulbert, *Moral Responsibility*
Deborah Perron Tollefsen, *Groups as Agents*
Joshua Weisberg, *Consciousness*
Chase Wrenn, *Truth*

Probability

Darrell P. Rowbottom

polity

First published in 2015 by Polity Press

Polity Press
65 Bridge Street
Cambridge CB2 1UR, UK

Polity Press
350 Main Street
Malden, MA 02148, USA

ISBN-13 978-0-7456-5256-6
ISBN-13 978-0-7456-5257-3 (pb)

A catalogue record for this book is available from the British Library.

Library of Congress Cataloging-in-Publication Data

Rowbottom, Darrell P., 1975–
Probability / Darrell Rowbottom.
pages cm
Includes bibliographical references and index.
ISBN 978-0-7456-5256-6 (hardback : alk. paper) – ISBN 978-0-7456-5257-3 (pb) 1. Probabilities–Philosophy. 2. Decision making–Philosophy. I. Title.
QA273.A35R69 2015
519.2–dc23
2015005729

Typeset in 10.5 on 12 pt Sabon
by Toppan Best-set Premedia Limited
Printed and bound in the UK by CPI Group (UK) Ltd, Croydon, CRO 4YY

For further information on Polity, visit our website:
politybooks.com

To my parents, Errol and Jean,
in gratitude for forty years of loving support

Contents

Preface

I have aimed to produce a highly accessible introduction to the philosophy of probability, which will be of interest to students in *any* discipline where probabilties are used. It doesn't just present the many different ways that probabilities may be interpreted, and the arguments for and against each, with reference to everyday life. It puts the findings to work too. It explains several common fallacies in probabilistic reasoning (in Chapter 9). It also covers the interpretation of probability in the social sciences and natural sciences (in Chapter 10).

Writing the book has been a labour of love, because most of the prestige in academia nowadays is associated with research productivity. As such, sadly, writing an introductory book isn't typically in one's own best interest (especially as a junior academic). (I began writing this book in 2010, when I was indeed a junior academic, and was supposed to finish it by 2012. I am a very naughty author!) It will be reward enough for me, though, if my passion for the subject comes across in what I've written, and inspires you to study it further (if you haven't already). If it does – or if you have any questions about anything in here – then let me know!

I have many debts of gratitude. First, I'd like to offer my sincere thanks to everyone who has commented on drafts of this book, or parts thereof: Chris Atkinson, Jenny Hung, William Peden, Mauricio Suárez, Jon Williamson, Jiji Zhang, previous students on my 'Probability and Scientific Method' course, and the anonymous reviewers appointed by Polity. Second, I'd like to express my gratitude to Donald Gillies, who inspired my interest in the philosophy of probability, and taught me much of what I know about it. Third, I must thank several editors at Polity – Emma Hutchinson, Sarah Lambert, and Pascal Porcheron – for their heroic patience in helping me to see this project through. Fourth, and finally, I am grateful to Sarah Dancy for her meticulous copyediting, and for removing many an unwarranted exclamation mark!

1 Probability: A Two-Faced Guide to Life?

1 Why Care about Probabilities?

A book on how to understand probability may not sound interesting; in fact, it *probably* doesn't sound interesting if you're not interested in maths. But if you don't understand probability, then you'll *probably* find yourself making some bad decisions. (Maybe it would pique your interest if I told you that I made a lot of money, from people who didn't know as much about probability as they should have, during my student days? More on this in Chapter 3.) Sometimes you'll act when you shouldn't, and other times you'll fail to act when you should. Don't take my word for it. Let's think about scenarios in which claims involving probabilities are relevant in everyday life.

Imagine you're intent on climbing a mountain, and you consult the weather forecast for the day. On the report, you see that the probability of precipitation in the relevant mountain range – or what is sometimes called the *chance* of rain – is just one in twenty, or 5 per cent. Should you take waterproof gear with you?

Obviously this will depend a little bit on context, so let's fill some of that out. Imagine you don't have any waterproof gear, and that it will be quite a hassle to get some, but that you don't want to get wet. Overall, you think that getting wet would be more unpleasant than going through the trouble of getting the gear; in an ideal world, however, you'd neither get the gear *nor* get wet. It's possible to assign a number to each possible outcome, a *utility*, to make this kind of discussion more precise. Let's avoid complicating things unnecessarily, though. We can instead *rank* the four possible outcomes in order of your preference: no gear and no rain (*best*), gear and rain (*2nd best*), gear and no rain (*3rd best*), and no gear and rain (*worst*). (It's always helpful in such scenarios to think about whether anything has been assumed which hasn't been explicitly mentioned. I encourage you to do this throughout the book. In this case, for example, 'gear and rain' has been ranked more highly than 'gear and no rain'. I did this because I figured you'd be a bit irritated at having the gear if it didn't rain; you'd be thinking 'I shouldn't have bothered to get this gear!' But perhaps I should have added this as a stipulation in presenting the context.)

The order of preference makes it clearer what's at stake in this hypothetical scenario. If you take the gear, you miss out on the best possible outcome. But you also protect yourself from the worst possible outcome (while giving yourself a shot at the second and third best outcomes). Now if the order of preference were the only information you had, your choice might depend only on your attitude towards risk; some people are more *risk-averse* than others. But you also know that no rain is much more probable than rain, which may affect your decision. And indeed it should affect your decision, as we will soon see, when 'much more probable' is *interpreted in some of the available ways*. Very roughly, we may capture why by saying that probability is often used as a measure of the *salience* of various possibilities.

Maybe you are still unconvinced that probabilities, so construed, are important. So imagine that you failed to rank possibilities by salience. You would treat any possibility you identified the same as any other. You would treat the possibility of a meteorite landing on your head, or of being accosted by a knife-wielding psychopath, as seriously as you would the possibility of rain. You would be worried about whether to wear a hard hat, a stab-proof vest, and so forth. In fact, with a little imagination, you'd be worried about so many possible fates that you'd be overwhelmed and confused. (Of course, inaction can be bad too. If you stay in, you might die in an earthquake. And so on!) The only upside would be that you'd be able to consider lots of good possibilities as well as the bad ones; of stumbling on a hidden cache of diamonds, of meeting a future partner, and so forth. But really, you'd have no way of proceeding other than guessing about what was best to do. Life would be a series of such guesses. And most of us don't treat life that way, in so far as we think that anyone who wears a hard hat at all times is crazy.

However, this still leaves us with the question of what, exactly, we can and should understand probability talk to reflect. And this is the main question that this book focuses on. One way of thinking about the issue is as follows. How can we satisfactorily *translate* a claim like 'The probability of rain in Hong Kong today is 0.5' into a *descriptive* claim that does not involve a mention of probability? (Note the use of 'descriptive'. We don't want the claim *just* to be a statement about how one should act. We want it to provide a *reason* for acting in a particular way.) Whether such a statement *should be* used to guide one's actions – and which actions, if so – depends on how this is done.

We will come to this below. But beware that there are all kinds of little tricks and subtleties that are easy to miss if you don't think carefully about probability statements. Consider again, for instance, the scenario discussed above. You know that the probability of precipitation in the relevant mountain range is one in twenty. But you only expect to climb one mountain in that range. So might the probability of rain *on the mountain you intend to climb* (or better still, on the route you intend to take) be different from one in twenty? Might it be lower? Might it be higher? Have a think about this before you read on.

2 The Two Faces of Probability

Imagine that I take a normal coin out of my pocket. What do you think the probability is that I will get a result of 'heads' when I flip it? As it happens, I have asked this question to students on numerous occasions. The following dialogue illustrates how one of these discussions – the best one! – developed:

DARRELL: What is the probability that I'll get 'heads' when I flip this coin?

STUDENT ONE: The probability is one half.

DARRELL: Does anyone disagree?

STUDENT TWO: I think we *don't know* what the probability is.

DARRELL: Really? Why's that?

STUDENT TWO: We would need to do an experiment to estimate the probability.

DARRELL: So what would you say to Student One?

STUDENT TWO: The coin – or your flipping of it – may be biased. We don't know if it is. So we have to do an experiment to find out if it is, and by how much if so.

DARRELL: OK. So you think that if I flip the coin repeatedly, and we record the *frequency* of heads results, then we'll be in a better position to estimate the probability of a heads result on a flip?

STUDENT TWO: Yes, that's right.

DARRELL: So on your view, the *real* probability would definitely become apparent if, by some kind of magic, we carried on this process forever. That is, if I flipped the coin infinitely many times and we had all the results?

STUDENT TWO: I suppose so, yes.

DARRELL: OK. How would you respond to this, Student One?

STUDENT ONE: Well it's true that the coin might be biased…

DARRELL: …by which you mean that the process of me flipping the coin might turn out to give one result with a higher frequency than the other?

STUDENT ONE: Sure, we can understand it that way. But even if we imagine that the coin is biased, we don't know *which* way it's biased. So it seems right to conclude that the probability of a heads result is the same as the probability of a tails result *given what we know*.

DARRELL: OK. So this means that there's a real difference in your views. Correct me if I'm wrong, Student Two, but on your view the probability may be different if I swap this coin for another one? That is, because the experimental setup will change?

STUDENT TWO: That's right.

DARRELL: But the probability on your account will stay the same if I change the coin, Student One?

STUDENT ONE: Yes, that's right. Well...it's right if you don't tell us anything more about – or I don't otherwise know anything more about – the coin.

DARRELL: OK. So you're saying that the correct answer to a 'What's the probability?' question like the one we're discussing is relative to the information available?

STUDENT ONE: That's right. If you just tell me you have a two-sided coin then all I really know is that there are two possible outcomes when you flip it.

DARRELL: So you conclude that each possibility is as probable as the other?

STUDENT ONE: Yes.

DARRELL: Why's that?

STUDENT ONE: There's no way to choose between them with the information I have. Each possibility is just as 'live' as the other.

DARRELL: Good. So Student One has what I call an *information-based* view, and Student Two has what I call a *world-based* view. Normally these are given different names – *epistemic* versus *objective* or *epistemic* versus *aleatory*, for example – but I think these are a bit more confusing.

In summary, it is possible to interpret probability talk in two main ways; either as saying something about the world (*world-based*), or as depending on one's state of information (*information-based*). You may think that these definitions are a little vague. They are meant to be. In fact, there are lots of different, more specific, interpretations of probability talk that fall into these two categories. We will be looking at these interpretations throughout the book, and I'll give a list of them at the end of this chapter. For the moment, though, let's just think about how the dialogue above might continue. Let's consider how one might start to argue that probability should be understood as world-based or information-based.

STUDENT TWO: Hold on. I have a question for Student One.

DARRELL: Go for it.

STUDENT TWO: Why didn't you also consider the possibility that the coin lands on its side?

STUDENT ONE: Good question. Because I've learned something about the behaviour of similar coins, I suppose.

STUDENT TWO: So couldn't we say that you have *learned* that similar coins are unbiased – or, as Darrell put it, that the frequency of heads will be one half, in the long run, when they are flipped?

STUDENT ONE: I don't think I have learned that!

DARRELL: What you can say, Student One, is that you took *some information* that you've learned from experience, and used that to narrow down the possibilities. Actually, I guess you also assumed that the coin would land when flipped?

STUDENT ONE: Of course!

DARRELL: Right. You took all the relevant information you had – some from your own previous experience – and thought about how that could be brought to bear on the question. You were concerned with whether that *information* has some kind of relationship to the claim that 'The coin will land on heads when Darrell flips it'.

STUDENT ONE: Yes. And I would have given 'heads' and 'tails' and 'side' the same probability each, if I lacked any information about coin flips beyond the possible outcomes thereof.

STUDENT TWO: I see! So Student One claims not to be considering whether the coin is biased, but instead to be reporting on to what extent her available information – her knowledge, perhaps – suggests that 'The coin will land on heads when flipped' is true.

DARRELL: Excellent. I can see this is going to be a great class!

You might get the impression that choosing between world-based and information-based views will be tricky, since both students have reasonable perspectives. This is a good impression, because there are widespread differences of opinion in the academic world about which way is right, even though, as you'll see, the debate about this has been going on for a *very* long time.

3 Monism or Pluralism

But must probability talk be viewed in just one way? In order to answer this, let's continue the dialogue and introduce a new character.

STUDENT THREE: I think that both Student One and Student Two have reasonable views. Can't they both be right?

DARRELL: Good question. Let's think about it. Can they both be right about the probability of the coin flip?

STUDENT THREE: I guess not. It can't be one half and also something else! But it's one half on the view of Student One, and could turn out to be something else on the view of Student Two.

DARRELL: You're right that the probability cannot have two values at the same time under any given interpretation. So if I say 'The probability of heads is *r*', *r* can only have one value when 'probability' is understood in any particular way...

STUDENT THREE: Right. I see. But maybe we can still interpret 'probability' in each way? So maybe there are two different kinds of probability that the coin will land on heads? A world-based one, and an information-based one, which may have different values?

DARRELL: Absolutely. The point is to be consistent, and to avoid the fallacy of equivocation, where the same term is used to mean different things in the same argumentative context.

STUDENT THREE: So there's *definitely* not any logical problem with saying that the probabilities in quantum mechanics should be understood as world-based, and the probabilities concerning coin flips should be understood as information-based?

DARRELL: Indeed there isn't. It's fine to apply different interpretations of probability in different contexts. Actually, there's nothing *logical* to stop you from using *only* a world-based view for discussions involving coin flips, and *only* an information-based view for discussions involving die rolls, for example. Although that would be a bit weird...

STUDENT ONE: Because those two kinds of situation are so similar?

DARRELL: Right. So it would be hard to see a reason for using only a world-based view when dealing with coins, and for using only an information-based view when dealing with dice.

STUDENT TWO: But if we were dealing with, say, quantum mechanics as compared to weather forecasting, then there might be an excellent reason to use a single different interpretation for the probabilities used in each?

DARRELL: Many philosophers who advocate *pluralism* – who think that more than one interpretation of probability is legitimate – would answer 'Yes'. Karl Popper, for example, thought that the probabilities in quantum mechanics are world-based, and that the probabilities involved in the confirmation of scientific theories are information-based. We'll discuss both these areas in Chapter 10.

STUDENT THREE: So what's the case for *monism* about probability?

STUDENT ONE: Simplicity, elegance, and unity may be considerations. Things might end up being a lot neater.

DARRELL: Yes, they might indeed. But we may not want to sacrifice explanatory power for simplicity. And, besides, perhaps the world is really a complicated place?

STUDENT ONE: I see what you're saying; that the simpler of two competing theories may not be the true one, or the truer one. But perhaps simplicity and unity should be used to choose between different accounts *when all other things are equal*?

DARRELL: Maybe. It's a tricky question to take in the abstract. Let's think about this some more when we look at the concrete proposal of Bruno De Finetti, who was a dedicated *monist* about one information-based view in particular. He struggled to *explain* why world-based views sometimes seem so appealing. Beforehand, we can also consider a classic argument for monism, due to Pierre-Simon Laplace.

We have seen that asking 'What is *the* proper interpretation of probability?' may be misguided. It may be wiser to ask instead, 'What are the proper interpretations of probability statements in such-and-such a context?' The choice is yours.

4 Laplace's Demon: A Thought Experiment

But how should you make the choice? One option is to use thought experiments. And Laplace, whom we will encounter again in the next chapter, used a striking example to argue for information-based monism. Here is an improved version of this.

Imagine a powerful being, a 'demon'. The following is true of it:

1 It knows everything about the initial state of our universe – it knows about all things that existed initially, and all the properties thereof.
2 It knows all the fundamental laws of nature in our universe, which govern how the things therein behave.
3 It has the ability to quickly perform any calculation, no matter how complex.
4 It is not part of our universe.

Now consider whether the demon requires probabilities, when it comes to predicting anything about our universe. Laplace's answer was a resounding 'No!' Or to put it more precisely, his conclusion was:

5 It is able to quickly determine the state of our universe at any point in time.

(Laplace's original thought experiment involved a demon that knew the present state of its own universe. And he claimed that it could work out all the future *and* past states. But this creates many problems that my version avoids.)

Let's try to get further insight into the demon's situation. Consider an extremely simple universe, composed of only two things, governed only by the laws of classical mechanics and gravitation. Imagine there are only two tiny indivisible (and perfectly solid) spheres, one of which orbits the other in a circle. Even we, with our limited intellects, can work out the state of such a universe at any point in time, provided we know the initial positions, velocities, and masses of the spheres (as well as the laws).

So Laplace's idea was that we need probabilities *owing to our ignorance*. We can see this by imagining that we lack one piece of information in the simple case above. Say we lack knowledge of the initial position of the non-orbiting sphere. Now we cannot predict the future states of the universe, at any specific time, precisely. However, we can still use probabilities to say *something* about those states.

But is the argument above valid? Is it impossible for the conclusion to be false, if the premises are true? No, because there is a crucial hidden assumption. (There may be others, but this one is especially important.) This assumption is that the fundamental laws of nature do not *contain* probabilities. But why should we think that? Why should we *assume* that there is only one possible future for any given initial state? The universe might instead be *indeterministic*; and we will return to this idea much later, in Chapter 8, when we consider Popper's propensity view of world-based probabilities.

It is worth adding that there are some areas of physics, such as quantum mechanics, where the best (candidate) laws we have *do* involve probabilities. So (even my improved version of) Laplace's thought experiment is not as good as it first seems.

5 Interpretations of Probability: An Initial Taxonomy

As I mentioned above, there are several different interpretations of probability that fall into the two overarching categories, world-based and information-based. They are as shown in table 1.1.

In the next part of the book, we will look at the information-based views. The part after that covers the world-based views. In the penultimate chapter of the book, when we're armed

TABLE 1.1 A taxonomy of interpretations of probability

Information-based interpretations	World-based interpretations
Classical	
Logical	
Subjective	Frequency
Objective Bayesian	Propensity
Group	

with these interpretations, we will look at some fallacies, puzzles, and paradoxes that arise in using probabilities. Finally, we will consider how probability bears on how we may understand a selection of theories from the humanities, the natural sciences, and the social sciences.

2
The Classical Interpretation

The mathematical theory of probability developed because of a pressing need at the gambling table. (Gambling will be a recurring theme in the book.) Patterns that weren't understood were spotted in results, especially in dice games, and it was natural to look for some kind of mathematical means of understanding (and better still, predicting) them. There was also a problem of how to divide money between players fairly when a series of games had to be stopped prematurely. If I have a better chance of winning than you, it would seem fair that I get more of the money put into the pot than you. But how could we measure our relative chances of winning? How should we divide the stakes?

Enter probability theory (or, at least, the first steps in developing the modern theory of probability) as a result of exchanges between three characters in seventeenth-century France: a gambler called Antoine Gombaud (a.k.a. Chevalier de Méré), and the mathematicians Blaise Pascal and Pierre de Fermat. The gambler asked the question about how to divide stakes – which was posed originally, more than 100 years earlier, by a mathematician called Luca Pacioli – and the others solved it. Here's a simplified version of the question.

Two people agree to play a series of *fair* games for money, and each puts the same stake into a pot. (It is important that the games are *fair*, as we will see. We can imagine these to be flips of an unbiased coin. A 'heads' result counts as a win for player one. A 'tails' result counts as a win for player two.) They agree that the first to win three games will collect the money in the pot. Unfortunately, however, they are forced to stop playing before either player has won three games. At the point at which they stop, player one has won two games, and player two has won one game. What to do with the money in the pot?

Pascal and Fermat recognized that the way to answer this question is to consider the future possibilities for the players, in the situation in which they stop. Others who tried to answer the problem typically just considered what had happened in the past at the stopping point, and tried to divide money on that basis. One suggestion, for example, was that player one should be given two-thirds of the pot, because he has won twice as many games as player two. But as we'll see, this isn't right.

Let's begin by listing the possible futures in which the series of games would be complete:

1 Player one wins the next game. (Therefore, player one wins the money. The final score will be 3–1.)
2 Player two wins the next game, but player one wins the game after that. (Therefore, player one wins the money. The final score will be 3–2.)
3 Player two wins the next two games. (Therefore, player two wins the money. The final score will be 2–3.)

Let's now move to dialogue format, to see where we can go from here.

DARRELL: Let's imagine that I reason as follows. There are three possible outcomes. Player one wins in two of those outcomes, whereas player two wins in the remaining outcome. Therefore two-thirds of the stakes should go to player one, and the remaining third should go to player two. What's wrong with this?

STUDENT ONE: Well the first possibility will happen half the time, because the coin is fair...

DARRELL: Right. Let me stop you there for a moment, to help others to follow. I think we can break this problem down into chunks. So we now know that at least half the money should go to player one?

STUDENT ONE: Yes. And I see what you're going to suggest next: that we can now think about what should happen to the rest of the money?

DARRELL: Quite right. Now we jump to thinking about the remaining two possibilities...

STUDENT TWO: Or to put it differently, we move to considering a situation where the scores are even, with two wins each, but the stake is half the total in our original scenario. And we ask how it would be fair to split that stake if the game were not completed.

STUDENT ONE: Very clever! And now you've made the answer to this part obvious: their scores are even, and the game is fair, so an equal split is fair. Yes?

STUDENT TWO: My thoughts exactly. So now we just sum the totals. Player one should get a half, plus a half of a half (which is, of course, a quarter). So player two should get one quarter, and the rest should go to player one.

DARRELL: Excellent reasoning. You considered the question in steps involving fair games, and the answer became evident. Of course, one can use a bit of mathematical trickery to avoid listing every outcome; but the basic strategy you've used is correct.

A diagram, depicting possible outcomes, will help to understand the reasoning used by the students in this dialogue. Let's represent the scores as (x, y), where x is the score of player one and y is the score of player two. The starting position is (2, 1), and the possible outcomes are (3, 1) and (2, 2). Then from (2, 2), the possible outcomes are (3, 2) and (2, 3). Next to the arrows, we can write what fraction of the time the outcome they point towards will occur (imagining the same kind of scenario were to be repeated forever). So the number next to the arrow from (2, 1) to (3, 1), for example, represents the fraction of the time that (3, 1) will be the next outcome after (2, 1); that is, when a game is played. Since the games to be played are defined as *fair*, we can take it that the number is the same for every arrow in the diagram; half the time player one will win, and half the time player two will win, in each game.

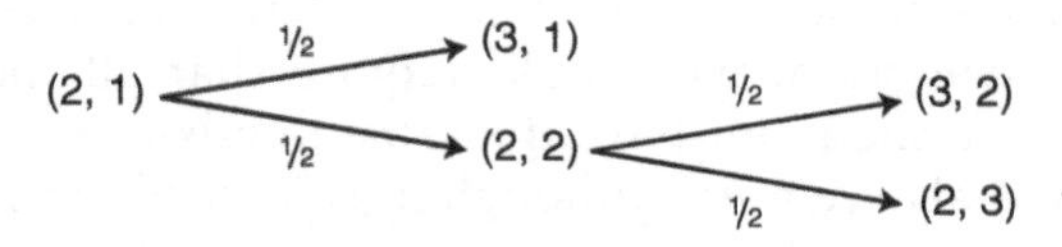

FIGURE 2.1 Possible outcomes in the problem of dividing stakes

Working out how often any outcome will happen from the starting point is now easy too. It's just a matter of multiplying the fractions next to the relevant arrows. So if we want to know how often the series of games will end in (2, 3) when the scores are (2, 1) – which is, in fact, how often player two will win in the scenario – we start at (2, 1) and follow two arrows, noting the numbers next to each. The first arrow is down to (2, 2) and the second is down to (2, 3). Each arrow is marked with a half. Now we multiply: ½ × ½ = ¼. We should expect (2, 3) to be the final result one quarter of the time. And since this is the only end state in which player two will win, it's now evident that giving player two one quarter of the pot, when the game ends at (2, 1), is fair.

We can now jump forward to the early nineteenth century, when Pierre-Simon Laplace gave the classic statement of the classical theory of probability:

> The theory of chance consists in reducing all the events of the same kind to a certain number of cases equally possible, that is to say, to such as we may be equally undecided about in regard to their existence, and in determining the number of cases favorable to the event whose probability is sought. The ratio of this number to that of all the cases possible is the measure of this probability, which is thus simply a fraction whose numerator is the number of favorable cases and whose denominator is the number of all the cases possible. (1814/1951: 6–7)

We may understand the talk of 'equally possible' and 'equally undecided about in regard to their existence' in terms of equal fractions of times that outcomes are *expected* to occur (on outcome diagrams like the one above). For example, we are equally undecided about whether player one will win any given game and whether player two will win any given game precisely because the game is defined as fair.

Now consider any point on such a diagram. According to Laplace's definition, we should require that all the arrows moving from the point have the same number next to them. (We should also require, although it isn't clear from the quotation above, that the numbers on the arrows from any given point add up to one. This is to ensure that all the relevant possible outcomes are on the diagram. To see this, imagine we replace all the halves with thirds in figure 2.1. Then we know that in any given game, player one will win a third of the time, and player two will win a third of the time. But this would mean that another third of the time, something else would happen which we hadn't included.)

Thinking in this way – in terms of the diagrams – makes it clear what's wrong with Laplace's definition. What happens if the game is not fair? What happens if it is biased in favour of one of the players, for example, because it involves a loaded die? Or what if it's a game of skill, and one player is better at the game than the other? Must we then remain silent about what the probabilities are? This would be a daft conclusion, when it is so obvious how we can alter diagrams like the one above in order to cope with biased situations. All we need to do is make the values on the arrows different, while ensuring that all numbers leading from any given point sum to one. Bingo! We can solve the problem of dividing stakes in unfair scenarios with no difficulty.

Consider the adjustment of figure 2.1, shown in figure 2.2, for instance. Now we are considering games that are biased in player one's favour. But we can still work out what fraction of the stakes should go to player two if a series of games is concluded at the (2, 1) point. As before, we multiply the fractions on the relevant (downward pointing) arrows. We discover that player two will only end up winning one-ninth of the time. So we know that a fair way of dividing the stakes

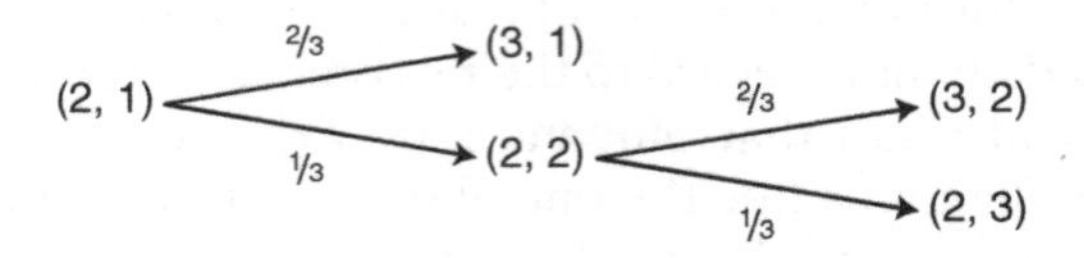

FIGURE 2.2 Possible outcomes in the problem of dividing stakes in an unfair game

is such that player two gets one-ninth, and player one gets the rest.

Before we press on to consider other options, however, we should consider a final, thought-provoking, dialogue concerning the discussion above:

STUDENT ONE: Hold on. Aren't the numbers next to the arrows actually *probabilities*?

DARRELL: Yes, they are. If you think about the rules for adding and multiplying them, some of which have been mentioned previously, this becomes clear.

STUDENT ONE: So didn't you, in the way you just set things up, assume that they were world-based? After all, you talked about 'equal fractions of times that outcomes will occur'.

DARRELL: You make an excellent point. I chose to do that to make the presentation clear. But I could have chosen to do things a bit differently. Think, for instance, about how else I might have defined a 'fair' game...

STUDENT TWO: How about as a game where we simply have no reason to expect one outcome rather than another?

DARRELL: That would do the trick! Then we are, to quote Laplace, 'equally undecided... in regard to' which possibility will occur – that's to say, as to which player will win – although we needn't assume that each will occur as often as the other in the long run.

STUDENT ONE: I see. And actually, now I think about it some more, I guess the probabilities used above – the values next to the arrows – may be information-based even if we think of fairness as you defined it.

DARRELL: Can you explain why?

STUDENT ONE: Sure. We can have no reason to expect one outcome rather than another *because* we know only that player one will win as often as player two in the long run.

DARRELL: That's exactly right. Well spotted. So we could use knowledge about probability in a world-based sense – or, if you prefer to insist that there are no world-based probabilities, just frequencies of events – to help our assignments of probability in an information-based sense. We will return to this topic later, in Chapter 5, when we discuss an interpretation of probability called 'objective Bayesianism'.

Further Reading

For more on the early history of probability theory, consult David (1962), Hacking (1975), and Daston (1988). Each of these books is intermediate to advanced in difficulty (considered section-by-section); David (1962) is the most accessible overall, although the other two texts are of higher academic quality.

3 The Logical Interpretation

The basic idea behind the logical interpretation, which was introduced by the economist John Maynard Keynes in 1921, is that there are logical relations between propositions (or statements) other than entailment relations. To understand what this means, let's first look at how we might define entailment relations in terms of probabilities. We will need to use *conditional* probabilities – such as $P(p, q)$ or $P(p|q)$ (which are two different popular ways of writing the same thing) – in order to do this. Roughly, $P(p, q)$ represents the probability of p given q. More precisely, as is explained below, it represents the probability of p on the assumption that q is true.

1 A Brief Introduction to Conditional Probabilities

An example will help to understand conditional probabilities. Imagine that I offer you a bet right now, at even odds (meaning that you'll give me a sum of money, and I'll give you double that if you win), that you won't read every chapter of this book. Imagine also that the sum of money you'll have to bet (your 'stake') if you accept my offer is enough to do something fun with – to have a great night out, say – but no more. (So if you win, you could, for instance, have two great nights out instead of one.) You might not want to take the bet, because you can't be sure you'll read all the chapters. And it wouldn't be worth committing to reading them all just to have an extra good night out.

However, now imagine I offer you a different bet. It's just like the other bet, but it's *conditional* on you reading up to the penultimate page of the last chapter. What this means is that the bet is only effective if 'You have read all of this book up to the penultimate page of the last chapter' becomes true. If it never becomes true, you'll not lose any money. (The bet will be 'off', as is sometimes said.) But if it becomes true, the bet will be on. At this point, you will give me your stake. If you then read the last page of the last chapter as well, you'll win double your stake. If you don't, then you'll lose your stake. I guess that this bet seems much more appealing to you. You might think that knowing you'd made the bet would encourage you to read just one more page, at that point. (I am sorry to tell you, though, that I am not *really* going to offer it to you!)

Now let p be 'You will read every chapter of this book' and q be 'You have read all of this book up to the penultimate page of the final chapter'. In the first case discussed above, we might say that your estimation of $P(p)$ – an *unconditional* probability – will affect whether you take the bet. In the second case, though, it will be your estimation of $P(p, q)$ that will count. (Actually, on a topic we'll return to shortly, we might think that both the probabilities you thought about were conditional. In the first case, you have some background assumptions you're making, some of which may turn out to be false; for example, you may have an expectation that the later chapters in the book will be as enjoyable to read as the first one. Let's represent all these background assumptions as b. Then we might say you *really* considered $P(p, b)$ in the first case. You might like to think about what you really considered in the second case, before reading on, too.)

2 What Are Logical Probabilities?

Now let's think about how to represent relations of logical entailment in terms of probability. Imagine q entails p (and

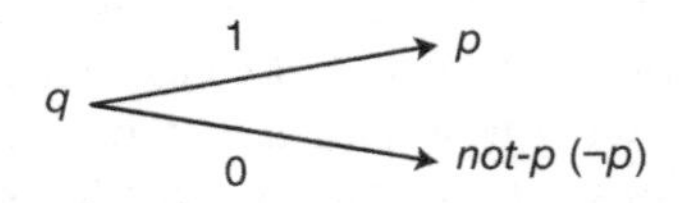

FIGURE 3.1 Outcome for p in logically possible worlds where q, when q entails p

that neither p nor q are contradictions). What value will P(p, q) take? The answer becomes clear if we think of entailment in terms of logical possibility (or logical necessity). For to say that q entails p is just to say that it is *not* (logically) possible for p to be false when q is true. And since there are only two possible values for any given proposition, namely true or false, we may conclude that P(p, q) is equal to one.

If this is still not clear, think back to the diagrams we used when discussing the classical interpretation. When considering P(p, q), the state we're interested in is q; q is the 'given'. And there are two possible values for p, in general; in each and every logically possible world, p is either true or false. So now, in order to put values next to the arrows on our diagram, we can ask in what fraction of logically possible worlds where q (the 'given') is true it is also the case that p is false. (A logically possible world is just a world where the laws of logic are not violated.) The answer, from our definition of entailment, is none. So zero must go next to the bottom arrow. And one must go next to the top arrow, by a process of elimination. ('It is not possible for p to be false when q is true' is equivalent to 'It is necessary for p to be true when q is true' when it's assumed that p may only be true or false.)

Similar reasoning shows that P(p, q) is zero when q contradicts p, on a logical view of probability. To see this, we need only note that contradicting p is the same as entailing *not-p* (which I will write as $\neg p$). And P($\neg p$, q) is one when q entails $\neg p$, so P(p, q) must be zero. In short, if $\neg p$ is true in all worlds where q is true, then p must be false in all worlds where q is true.

A natural question to ask next is 'What's P(p, q) when q is irrelevant to whether p?' Let p be 'The summit of the highest mountain in England is 978 metres above sea level' and q be 'Darrell has blue eyes'. These are (plausibly) completely unrelated, although true; q is not evidence for p, and p is not evidence for q. Let's now consider just those logically possible worlds where q is true. Should we expect that in most of these, p is true? It would seem not. Should we expect that in most of these, p is false? Again, it would seem not. So it seems fair to conclude that $P(p, q)$ is a half. This is the only remaining option. (In general, as you can see in Appendix A, $P(p, q)$ is equal to $P(p)$ when p is independent of q. So $P(p, q)$ will only be one half, when p and q are independent, on the condition that $P(p)$ is one half. We'll cover how to represent unconditional probabilities, like $P(p)$, in the next section.)

What's left? Cases where, to use the phrase introduced by Rudolf Carnap (1950), *q partially entails p*. If q partially entails p more than $\neg p$, $P(p, q)$ will be greater than one half but less than one. And if q partially entails $\neg p$ more than p, $P(p, q)$ will be less than one half but greater than zero. In either situation, recall, $P(\neg p, q)$ will be $1 - P(p, q)$. Or to put it differently, $P(\neg p, q) + P(p, q) = 1$ because either p or *not-p* must be true.

Here's an example where it appears that $P(p, q)$ is much greater than one half:

q: 1000 sexually active patients took a pill of type A, every day for a year. None of these patients became pregnant during that year.
p: Mary will not become pregnant during a period in which she takes a pill of type A every day.

The first piece of information, q, appears to suggest that pills of type A are contraceptives; and we can imagine getting such information from a medical trial of the pills. (Imagine we've ruled out people who got pregnant but forgot to take the pill at some point.) However, it is important to note that q does not say anything about a medical trial, and that new information could come along which would suggest that p is doubtful. Imagine, for instance, you discover that:

r: All the patients referred to in q were men.

Now there does not appear to be any evidence for or against p. And one philosopher of science, Peter Achinstein (1995), makes an even more radical suggestion. He thinks that the discovery of r would show us that $P(p, q)$ was not, as we initially thought, much greater than one half. On his view, the extent to which q is evidence for p is not a *logical* matter at all. Instead, it is a matter for empirical study whether q supports p (or supports $\neg p$). However, the natural way for an advocate of the logical view to respond – or, at least, to begin to respond – is to say that we must be wary of confusing $P(p, q)$ with $P(p, q \,\&\, r)$. The value of $P(p, q \,\&\, r)$ doesn't necessarily tell us anything about the value of $P(p, q)$. When we get new information, we consider different conditional probabilities to be relevant to working out whether p is true. That's all.

3 Conditional and Unconditional Probabilities in the Logical Interpretation

On the logical interpretation, it makes little sense to think of probabilities as being genuinely standalone, or unconditional. So to talk of some proposition being probable – for example, 'China will *probably* be the greatest economic power in the world by 2020' – does not make any sense, when taken literally. This is no surprise, given the underlying idea that probabilities represent partial degrees of entailment. To say 'p is partially entailed' raises the question 'By what?', in exactly the same way that 'p is entailed' does. Keynes expresses this idea elegantly:

> No proposition is in itself either probable or improbable, just as no place can be intrinsically distant; and the probability of the same statement varies with the evidence presented, which is, as it were, its origin of reference. It is as useless...to say '*b* is probable' as it would be to say '*b* is equal', or '*b* is greater than',...(1921: 6–7)

When thinking in terms of logical probabilities, we must therefore be mindful of precisely what is taken as 'given', or understood to lie on the right-hand side of the comma in the formula for the conditional probability under discussion. When someone (truthfully) says 'I will probably come' regarding an event, like a philosophy seminar, they are typically working on the basis of their relevant personal background information. If they later said 'I probably won't be able to make it', this would normally be because their background information had changed. They might have learned something new, e.g. that they'd become ill or that there was the possibility of a hot date instead. (Hot dates are usually better than philosophy seminars. Trust me.)

However, there is nothing to stop us defining unconditional probabilities in terms of conditional probabilities. One nice trick suggested by Popper, for example, is to define the unconditional logical probability of p as the logical probability of p conditional on any tautology, T. (Examples of tautologies, for those unfamiliar with logic, are $\neg(p \,\&\, \neg p)$ or 'It is not the case that p and *not-p* are true' and $p \vee \neg p$ or 'Either p is true or *not-p* is true'. These are called *the law of non-contradiction* and *the law of the excluded middle*, respectively. They are true in all logically possible worlds.) In short, Popper said that $P(p)$ should be understood to represent $P(p, T)$. That's no objection to writing '$P(p, T)$' as '$P(p)$' in so far as the mathematics is concerned.

4 Logical Probabilities and Beliefs

Before we press on, let's pause for a moment to think about how logic relates to beliefs. This is worth doing because some people talk about the logical interpretation as if it concerns only what we should believe, although this is misleading. Even Keynes appears to make this mistake at one point, when he writes:

> Let our premisses consist of any set of propositions h, and our conclusion consist of any set of propositions a, then, if a knowledge of h justifies a rational belief in a of degree α, we say that there is a *probability-relation* of degree α between a and h. (1921: 4)

The language here is rather different from our earlier talk of logical relations, such as entailment and partial entailment. But in fact, as Keynes explains a little later, his idea is simply that logical relationships constrain what it is reasonable for us to believe:

> ['Probability'] in its most fundamental sense...refers to the logical relation between two sets of propositions... Derivative from this sense, we have the sense in which...the term *probable* is applied to the degrees of rational belief. (1921: 11)

To see the basic idea, think again of entailment relations. If p entails q, but I believe p and $\neg q$, then I must, on Keynes's view, have *irrational* degrees of belief. Why? Because I've failed to recognize that it is *impossible* for p to be true and q to be false.

What are the 'degrees of rational belief' that Keynes writes of? We will discuss these in depth in the next chapter. For the moment, just think of these as 'degrees of rational confidence'. So if you know p, and p entails q, then you are rational to be totally confident that q (if you consider it); if you are confident that q to a lesser degree, you fall short of being rational. Similarly, if p partially entails q to degree r, and you know p, then you should not be confident that q to any degree other than r. That's Keynes's view.

Note well, however, that one could adopt a logical interpretation of probability and *disagree* with Keynes's account of degrees of rational belief. For example, you might think it is sometimes rational to believe in something for which you have no evidence (if, for example, there will be pragmatic benefits). Pascal's wager provides a good example. Roughly, it goes like this. If you believe in God, you'll get amazing benefits if He does exist (e.g. entry to heaven) and lose nothing if He doesn't. If you don't believe in God, you will incur a terrible penalty if He does exist (e.g. eternal damnation) and will otherwise gain nothing. So you should believe in God. If you *could* just choose to believe in God – it is plausible that you can't – then it would be worth taking this line of argument seriously. (There may be other problems with the argument – e.g. you may lose something if you believe and you're wrong. You might, for example, spend a lot of time in church that would be better spent elsewhere. But the wager nevertheless suggests that belief may be reasonable for purely pragmatic grounds in *some* circumstances.)

Alternatively, somewhat less radically, you might think that Keynes is just a little too strict, and that degrees of confidence should only be *approximately equal* to degrees of partial entailment, or something like that. The possibilities are endless, but let's not dwell on them. As this chapter concerns the logical interpretation of probability, our focus should be on the (alleged) logical relations that this rests on.

5 Measuring Logical Probabilities

We have now covered the fundamentals of the logical view of probabilities, and determined the values of logical probabilities in some special circumstances (e.g. when entailment relations are present). But we are left with the question of how we should calculate values for logical probabilities when partial entailment relations are present. Think back to the scenario above involving Mary, for example. What is supposed to be the *precise* value for $P(p, q)$, if anything?

In fact, answering this question is extremely difficult. So we will instead have to settle for covering how we may approach measurement in more general terms. We'll cover the account offered by the hero of this chapter, namely Keynes, which has proven to be one of the most influential and is also one of the most careful and systematic.

Keynes begins his account by suggesting that we can recognize some probability relations on the basis of intuition, or insight. And in saying this, he appears to have in mind that we have some kind of extra-sensory ability, or faculty, for grasping relations between propositions. An immediate response might be that this seems a little mystical. But let's explore this further in a short dialogue.

STUDENT TWO: I don't think I have this kind of ability! I'm pretty convinced that all my knowledge comes from experience...

STUDENT ONE: But surely Keynes didn't have to deny that?

DARRELL: Can you explain why you think that?

STUDENT ONE: Well for one thing, to say that we can grasp relations isn't to say we can understand propositions – get what they mean, for want of a better expression – without experience. Take 'The sky is red' as a case in point. Without experience, we wouldn't understand it.

STUDENT TWO: OK, that seems right. So Keynes could be an empiricist up to a point...

DARRELL: And so he was, up to *at least* that point...

STUDENT ONE: And why not even more? Let's think about the basis of non-logical knowledge. Insight may not, on this account, let us know a reasonably basic fact like 'I can see a red thing right now'. That could be a matter for experience.

DARRELL: Right. So one could think, as I believe Keynes did, that we start from an empirical foundation of what you called 'basic facts', or what I'd call 'observation statements', and can then work our way up to greater knowledge by our rational insight. This would be an ability to grasp relations between theories and actual 'observation statements', theories and possible 'observation statements', and even different (possible or actual) 'observation statements'.

STUDENT TWO: Can you give an example?

STUDENT ONE: I can. We may recognize that 'There is a non-white swan' falsifies the theory 'All swans are white' once experience has helped us to grasp what 'white' and 'swan' mean. But only experience can tell us whether 'There is a non-white swan' is true, and therefore whether 'All swans are white' is false.

DARRELL: Indeed. That would be a relatively uncontroversial example, as it involves entailment. Keynes just thinks there are similar cases involving partial entailment.

In fact, Keynes wrote: 'If the truth of some propositions, and the validity of some arguments, could not be recognised directly, we could make no progress' (1921: 53, f.1). But he did not think that *all* probability relations can be recognized directly. Quite to the contrary, he thought that we often need to *calculate* probability relations, and can use a special principle, namely the Principle of Indifference, to do so.

The same is true for normal entailment relations, on Keynes's view. We can spot some easily, e.g. that 'Tim is a black rabbit' entails 'Tim is a rabbit' (or to put it more for-

TABLE 3.1 Truth table for $p \oplus q$ and $\neg(p \leftrightarrow q)$

p	q	$p \oplus q$	$p \leftrightarrow q$	$\neg(p \leftrightarrow q)$
T	T	F	T	F
F	T	T	F	T
T	F	T	F	T
F	F	F	T	F

mally, p & q entails q). But other such relations we have to work out, to continue the quotation from the previous paragraph, by using 'the method of logical proof... [which] enables us to know propositions to be true, which are altogether beyond the reach of our direct insight'. For example, we might need to use a truth table to determine if an entailment relation is present in more complicated cases.

Here's an example. Think about whether 'It is not true that: p if and only if q' entails 'Either p or q'. Table 3.1 enables us to determine the answer. (Here's an explanation of the logical jargon, if you are not familiar with it: '$\oplus$' stands for 'or' in the exclusive sense of 'either... or' where 'both' is ruled out, and '$\leftrightarrow$' stands for 'if and only if'. Thus '$p \oplus q$' means 'Either p or q (but not both p and q)' and '$\neg(p \leftrightarrow q)$' means 'It is not true that: p if and only if q'.) Each row – rows run horizontally – represents a possible set of values for all the statements. The four rows, taken together, exhaust all the possibilities. Consider the first row. It shows that when p is T(rue) and q is T(rue), both the statements we're interested in – $p \oplus q$ and $\neg(p \leftrightarrow q)$ – are F(alse). Then we move on to look at the situation when p is F(alse) and q is T(rue). And so on.

We see from this table that $p \oplus q$ is true whenever $\neg(p \leftrightarrow q)$ is true, and vice versa. (There are two possibilities, in the second and third rows, where both are true. Neither is true in the other two rows.) So in fact, these statements entail *each other*. More than this, we see that each is false when the other is false. (In fact, the columns under each are identical. This shows they have the same values in all possible situations.) Thus they have a stronger relation still; they are *logically equivalent*, or say the same thing. But without training in logic, this is far from obvious. And it is not difficult to construct much more complicated relations than the one above, which even expert logicians would not be able to recognize without using a truth table, or some other method of proof. That's the point of the example: you can't spot the entailment relations without working through a proof. So don't fret if you find the details difficult, because you haven't studied logic.

Now we cannot use truth tables, such as the one above, to determine what kind of relations of partial entailment hold between two statements (if any). Rather, we need the special principle mentioned previously. Keynes describes it as follows:

> The Principle of Indifference asserts that if there is no known reason for predicating of our subject one rather than another of several alternatives, then relatively to such knowledge the assertions of each of these alternatives have an equal probability. Thus equal probabilities must be assigned to each of several arguments, if there is an absence of positive ground for assigning unequal ones. (1921: 42)

Imagine that I am going to choose a whole number between one and ten. What is the probability that I will choose five? Here, you have no reason to think that I will choose any particular number rather than any other. So you should assign an equal probability to each possible outcome, according to the principle of indifference. Since there are ten possible outcomes in total, the probability of each will be one-tenth. Easy. Much easier, indeed, than doing a standard logical proof. Or so it first seems. (This approach is also intuitively plausible. I have heard clever and highly educated people, such as academic colleagues, say things like 'I have a quarter of a chance of getting the job!' because they've made it onto a shortlist of four for a job interview. But is that right?)

6 Problems with the Logical Interpretation

The most serious problem with the logical interpretation is the issue we've just tackled, namely of how exactly we are to measure logical probabilities. If we can't measure logical probabilities as we've defined them, then we may begin to doubt that they exist at all. In particular, we may doubt that there are relations of partial entailment that come in degrees, or indeed any relations of partial entailment at all, between propositions.

The basic problem is that the principle of indifference does not tell us how to carve up the possibilities. This is nicely illustrated by an episode that happened in my student days. (This is a true story. I had a misspent youth!) I had recently attended a lecture on entropy and statistical mechanics – in the way the lecturer put it, I recall, on how disordered situations are more probable than ordered ones – and I was busy bragging about this in a pub. Somehow I got into a dispute with another regular drinker there, a wealthy lawyer, and I ended up making a series of bets with him in order to settle it. What I proposed – and I took the idea of using coins straight from the lecture – was that we'd flip five coins, and that I would win £5 if the result was two heads or three heads, whereas he'd win £5 otherwise. I also said we could continue doing this until one of us wanted to stop. He eagerly accepted the bet. And he began to lose. But he was convinced that I was just being lucky, and continued for over an hour (until the pub closed). By the end, I'd won around £60 and had been bought several drinks by amused onlookers. Not bad for a night's 'work'! The lawyer left, convinced that he had just been unlucky, despite my protestations, and challenged me to continue the next time we met. But I felt a little guilty about continuing, so I politely refused. (My only loss from the episode was a mild hangover the next day.)

What happened? The simple answer is that the lawyer assigned equiprobability to what Keynes called *divisible* outcomes, whereas I opted for *indivisible* outcomes instead. That's to say, the lawyer considered the following:

5 heads	Possibility One
4 heads	Possibility Two
3 heads	Possibility Three
2 heads	Possibility Four
1 head	Possibility Five
0 heads	Possibility Six

Because the lawyer assigned each of these outcomes the same probability, namely one in six, he expected me to lose more often than not. More specifically, he thought I would only win on two outcomes out of a possible six, and therefore only one-third of the time. No doubt he was looking forward to silencing the precocious young rake before him!

From my perspective, however, I was on to a winner. This is because I considered *indivisible* outcomes instead, and assigned these equal probabilities. (At least, these outcomes are indivisible if a game was only to be considered completed when each coin had landed on tails or heads, so that any event such as a coin landing on its side was to be discounted. These were implicit rules.) To be more specific, I was mindful of the *ways in which* each of the possibilities enumerated above might occur. Consider the following, where H represents a 'Heads' result and T represents a 'Tails' result, to see how I was thinking:

HHHHH   Five Heads Occur on One Possible Outcome

HHHHT
HHHTH
HHTHH   Four Heads Occur on Five Possible Outcomes
HTHHH
THHHH

TTHHH
THTHH
THHTH
THHHT
HTTHH   Three Heads Occur on Ten Possible Outcomes
HTHTH
HTHHT
HHTTH
HHTHT
HHHTT

Symmetrically, as you will see if you imagine each H swapping places with a T (and vice versa) in the above, there are ten possible outcomes corresponding to a 'two heads' result, there are five possible outcomes corresponding to a 'one head' result, and there is only one possible outcome corresponding to a 'no heads' result. (In mathematics, an outcome such as 'two heads and three tails' is called a *combination*. 'HHTTT' is one *permutation* of that combination. To put it simply, order matters for permutations but not for combinations.)

Thus, from my perspective, the probability of 'two heads' or 'three heads' in the game was five over eight, and the odds selected were stacked in my favour. The fact that I won so much money suggests that I was right, although it is indeed *possible* that I was lucky. (Actually it is possible to calculate the probability of my victory on the assumption that the lawyer was right about the probabilities; to do this we would need to know exactly how many games we played, and the result of each. Unsurprisingly, since quite a bit of beer was involved, I don't remember.) If you ever meet me and want to help me to test this further, you are most welcome to play this game with me. For money, of course! If you have any rich friends who would like to play too, please bring them with you. The more the merrier.

Keynes would have said that I did the right thing, by selecting indivisible outcomes. Otherwise, we would arrive at the paradoxical situation where the Principle of Indifference would say that my probability of winning, in a given game, had two different values simultaneously: one-third (the result from lawyer's calculation) and five-eighths (the result from my calculation). You might think we could resolve this difficulty by instead suggesting that the lawyer's way was right, and my way was wrong. The problem about saying this, though, is that there are often different, incompatible, *divisible* ways of carving up possibilities. Imagine I am going to buy a rabbit, and ask you what the probability is that I will buy a black one. You could take 'black' and 'not-black' to be the two possibilities. Or you could take 'black', 'brown', and 'neither brown nor black' to be the three possibilities. And so on. You would get a different probability, by using the Principle of Indifference, in answer to one and the same question.

You might also be tempted to conclude that Keynes was right because I won. But this would be a mistake. After all, the coin (or coin-flipping procedure) might have been unfair. Or maybe I just guessed at the right *world-based* probability (by using the Principle of Indifference as a heuristic device). (Maybe I would have won eight-fifths of an infinite series of games like this.) In short, maybe I just got lucky by assigning equal probabilities to these possibilities, in this particular case. If I did the same thing with some other possibilities, like rolls of a loaded die, that might be a terrible mistake! And then I would probably be the loser.

So does appeal to indivisibility solve the problem about how to carve up the possibilities? Apparently not, because it doesn't deal with cases where the possibilities are infinite, or where there is no unique indivisible set of possibilities. This is nicely illustrated by the Horizon paradox, which is one of several such paradoxes that were presented by the mathematician Joseph Bertrand about thirty years before Keynes developed his logical view. Here it is:

> Imagine any plane in space; call this the Horizon. Now imagine another intersecting plane is chosen at random. What is the probability that the angle it makes with the Horizon is less than 45 degrees?

(This is a slightly different version from Bertrand's: he says choose any plane at random. This leaves infinitely many planes parallel to the Horizon, however, which just creates more confusion; and, indeed, he appears to ignore these planes in his calculations. Thus, I use 'intersecting plane'.) Let the angle be denoted by θ. This must be greater than zero but less than or equal to 90. So one could take each value to be equally as possible. But why not consider $\cos(\theta)$ instead, and consider each value of this function to be equipossible? Let's think about this.

STUDENT ONE: Using θ is more natural, isn't it?

STUDENT TWO: So it may seem to you, but isn't that just because of how you – well, all of us – happened to learn maths?

STUDENT ONE: That's a good point, I suppose; there's an element of convention.

DARRELL: It is. Actually, Bertrand gave an argument for using the cosine too; but we don't really need to get into that...

STUDENT ONE: Because the real worry is whether there are natural – actually, I guess, natural *and* unique – measures in all, or at least most, such paradoxical cases?

DARRELL: Quite.

STUDENT TWO: I was reading about another one – the water/wine paradox – yesterday. Here's how it goes. We have some liquid. We know only that it's composed entirely of water and wine, and that there is, at most, three times as much of one component as of the other. What's the probability that the ratio of water to wine is less than or equal to two?

STUDENT ONE: Is the idea that you can consider either the ratio of water to wine or the ratio of wine to water, and that you get different answers each way?

STUDENT TWO: Yes! So which ratio is the more natural?

STUDENT ONE: Nice. There really isn't one.

DARRELL: That's right. But I did read a paper a while back, by Jeff Mikkelson (2004), which argued that we should answer the question by thinking in terms of quantities rather than ratios. His basic idea is that the quantities are primary, and determine the ratios. Or more precisely: (a) a change in quantities is responsible for a change in ratios, and (b) if no quantities, then no ratios.

STUDENT TWO: I'm not sure I get it. How do we do a calculation now?

DARRELL: Mikkelson invites us to imagine the two components don't mix – just like crude oil and water – and to think of where the line separating them would fall if we poured the liquid into a measuring cylinder.

STUDENT ONE: I get it. So the answer would be the same no matter what volume of liquid we have in total. We don't need to specify what the scale on the measuring cylinder is!

DARRELL: Yes, it's very clever.

STUDENT TWO: But is it a unique solution to the paradox?

DARRELL: Actually, it isn't. Basically, Mikkelson chooses to discuss a variable that will have the same value if it's defined in terms of either ratio. But it is not the only such variable. There are infinitely many.

STUDENT TWO: But maybe, just maybe, it's the only *natural* solution?

STUDENT ONE: To be honest, I'm now not so confident that it's clear what 'natural' means. Here, I suppose it's the simplest solution, in addition to making sense from a physical perspective...

DARRELL: Even if the approach is the only natural one, there are still significant worries. Ask yourself if Mikkelson sticks to the information given in the question, or if he instead considers a different probability relation to the one asked about in the question *by importing some of his background information*. Or to put it differently, can we formulate his answer if we know *only* that there is wine and water in the liquid, obeying the stated ratio-based constraint?

STUDENT TWO: I guess we have to allow that we know some things more, strictly speaking – like maths, logic, and so on?

DARRELL: Yes, that's right.

STUDENT ONE: But those are rather different from all the physical information that Mikkelson uses.

DARRELL: Absolutely. That physical information really *doesn't* seem to be part of the question. Now maybe we can make some kind of move in Mikkelson's favour if we accept that words like 'wine' and 'water' imply dispositions; so that to know something is water is to know that it's disposed to behave in such and such a way in such and such a context.

STUDENT ONE: OK, but it does seem weird to suggest that, in order to know some stuff falls into such a category, I must know it possesses a long list of dispositions.

DARRELL: Right. Water has dispositions we didn't know about in medieval times, if we take science at face value. But it would be weird to conclude from this that medieval people didn't ever know when water was present.

STUDENT ONE: OK. So a physical assumption Mikkelson makes that isn't in the question is that the volume of the mixture is the sum of the volumes that the two components would each occupy if they were separated.

DARRELL: Well spotted! (It might not be. There could be some interaction.) And actually, there's nothing to stop us not using volumes at all. We might interpret 'as much of' in terms of something other than volume, like mass.

STUDENT TWO: So what's the upshot?

STUDENT ONE: The point, I suppose, is that Mikkelson is giving the scenario his own spin. And that there's nothing *in the question* to suggest that we should give it his spin, rather than another one.

DARRELL: I quite agree. We'll come back to this when we discuss objective Bayesianism, which can be seen as a successor to the logical interpretation, in Chapter 5.

In closing, it is worth noting that a kind of negative Principle of Indifference nonetheless remains plausible. As Keynes put it: 'two propositions cannot be equally probable, so long as there is any ground for discriminating between them' (1921: 51). Unfortunately this negative principle, taken alone, isn't sufficient to say what the values are (rather than are not) in any given case.

7 Partial Entailment vs. Partial Content

In my presentation of logical probabilities, I used the idea of partial entailment; so if *q* is evidence for *p*, then *q* entails *p* to some degree. But there is another way of thinking of logical probabilities, which is subtly different. This is to think in terms of content. Popper, in particular, advocated this approach at some points.

Think again of deductive arguments, and entailment. It is often said that if *p* entails *q* then the content of *q* does not go beyond that of *p*. Consider, as a case in point, an argument with *r* as the conclusion and *p* & *r* as the premise. Here, there is more information in the premise than there is in the conclusion; and at most, in a valid argument, there can be the same information in the premises as there is in the conclusion. Take '*p*, therefore *p*' as a case in point.

But when we consider non-deductive inferences, we find that this situation is reversed. The content of the conclusion is always greater than the content of the premises. Think of: '99 per cent of rabbits are brown. Tim is a rabbit. Therefore, Tim is brown.' The conclusion says something about Tim that is not mentioned in the premises, and therefore has greater content.

However, Popper's suggestion is that the premise '99 per cent of rabbits are brown and Tim is a rabbit' contains the content of 'Tim is brown' to a specific degree, which may be understood to be the probability of the latter given the former. In fact, when Popper (1983: 293) considers a structurally identical example – '92 per cent of men are mortal. Socrates is a man. Therefore, Socrates is mortal' – he says that the probability is (or is close to) 0.92.

This alternative is included mainly for the sake of completeness. It does not appear to have any impact on solving the measurement problem discussed above.

Further Reading

Most of the literature on the logical interpretation is advanced in character. However, Gillies (2000: ch. 3) provides a clear intermediate level introduction. Keynes (1921) is also highly readable.

4
The Subjective Interpretation

In the last chapter, we touched on the idea that beliefs can come in degrees; and we understood this, roughly, to mean that a person can be more confident about some things than others. I believe that I have written a textbook. I also believe that you will find some of the book enjoyable. But I am considerably more confident that I've written a textbook than I am that you, a reader picked at random, will find some of it enjoyable. So I have a higher degree of belief in the former than in the latter, although I believe in both. And these differences can exist even when we're comparing two things believed *not* to be the case. For example, I believe that President Obama will not be assassinated during the remainder of his term in office. I believe much more strongly, however, that one plus one is not equal to three.

Now the basic idea behind the subjective interpretation, slightly different versions of which were proposed independently by Bruno De Finetti (1937) and Frank Ramsey (1926), is that our degrees of belief should be constrained in particular ways if they are to be rational, and that those ways just happen to correspond to the axioms of probability. This seems surprising, at first sight, but it can be argued for in an elegant and simple way, namely by thinking about betting behaviour.

1 Dutch Books and Gambling Behaviour

Imagine we're going to have a bet together, you and I, about whether something will happen. It could be about whether your favourite sportsperson or team will win their next match, or something more trivial like whether it will rain tomorrow anywhere in your country. We then go on to agree on a stake, S, which is the maximum amount of money that could change hands.

I will choose whether to bet for or against the event happening. (Note that my talk of 'events happening' is just for convenience, and can easily be translated into talk of 'statements being true'. So we can understand the bets to concern propositions or statements. For example, betting on Manchester United winning their next match is equivalent to betting that 'Manchester United will win their next match' is true.) But before I make that choice, you're going to pick a number, a betting quotient b, on the understanding that the bet will be conducted as follows:

1 If I bet against the event occurring, you will pay me bS. If the event occurs, I will give you S.
2 If I bet for the event occurring, you will pay me $(1 - b)S$. If the event doesn't occur, I will give you S.

To choose b is to determine 'the odds' of the bet, which is usually expressed as a ratio, namely $b/(1 - b)$. To select a value of one half for b, for example, would be to give 'even odds' on the event; you'll double your initial payment if you win the bet, otherwise you'll lose it. (You may notice that if you select a value of one for b, there is no value for the odds when they're defined in the way above. This is an intended feature of the setup, as we'll see.)

We also want to set up the scenario so that the odds you give are fair, in your opinion, so there are a number of requirements that we should add. First, you're not to have any information about whether I intend to bet for, or against, the event occurring. If you do, you might stack the odds in (what you perceive to be) your favour. Imagine you know I'm going to bet that the result of a die roll is five, and you think that the chance of a five is one in six. You could choose to select odds as if getting a five were much more likely than that, e.g. even odds (as discussed above). Then I will only double my money if I win, but chances are – in fact in five times out of six, in your view – you'll just keep the money. Good for you! Bad for me! This is clear if you think about what you would expect to happen in a sequence of bets of this kind, with me continually betting on a five result.

A more homely example, which doesn't involve chances at all, can also illustrate the point. Imagine you are a car salesman. If I tell you that I want to buy a particular kind of car, you'll give me one price. If I tell you that I want to sell a car of the same kind, you'll give me a different (lower) price. So to elicit from you what you *really* think a car is worth, i.e. what a fair price is, I should refuse to reveal whether I want to buy or sell. I should just give you the information about the car and ask the value. The same is true when asking about odds, rather than monetary values; to specify the event and ask for fair odds, without saying whether the bet will be for or against the event occurring, is the way to go.

Second, S should be a set at a level that makes the bet worthwhile for you. By 'worthwhile', I mean not so high that you'd hate to lose it, or so low that you wouldn't care if you lost it. I guess for many student readers, something between five and twenty US dollars would be OK. But one hundred US dollars would be too high, and one US cent would be too low. If you're scared of losing the money, you might select unfair odds to try to protect yourself from losing too much. (If you set b at one half, then the most you can lose is half of S. Any other choice of b exposes you to a higher possible loss.) And if you don't care about losing the money, you simply don't care about selecting fair odds. There is no incentive to do so.

We could add other constraints. For example, you should not have control (or influence) over whether the event bet on occurs. Because then you could act to make it happen (or increase the chance of it happening) if I bet against, and could act to make it not happen (or increase the chance of it not happening) if I bet for. Moreover, we should also make sure that the bet is going to be concluded in the reasonably near future, so that you're not trying to minimize the amount of money you hand over. If you think hard, you may come up with other refinements you think are important. But let's not dwell on these. For the moment, let's simply recognize that when such details are worked out, implementing this kind of betting scenario will be much more difficult than it may initially seem. This is a problem we'll return to.

OK. We have our betting scenario. So let's now think about how you should – or, more importantly, how you *shouldn't* – assign values to b. To kick off, let's imagine you choose a value for b that is higher than one. Now if I bet against the event occurring, I will win money whatever happens. This is called making a *Dutch Book* against you. You will give me more than S, and at worst I will only have to give you S back. Thus, we can conclude that b should be no higher than one. (You may be curious about how to interpret having to 'pay' a negative value. The straightforward answer is that paying a negative amount involves being *paid* that amount; so to say that you would pay me $-S$ is to say that I would pay you S.)

And what if you choose a negative value for b? Now I can make a Dutch Book against you just by betting for the event occurring instead. You will pay me *more* than S, and I will only be obliged to give you S back if the event doesn't occur. Thus, we can conclude that b should not be lower than zero either. Overall, we have so far learned that $0 \leq b \leq 1$ for a sensible – or what we might call a 'rational' – bet in these circumstances.

There's more. Imagine now that you're completely convinced that what you're betting on will occur. For argument's sake, let's say it's a bet on something silly like 'Either it will rain tomorrow somewhere on Earth or it will not rain tomorrow somewhere on Earth'. (It's not logically possible for this to be false: it can't rain and not rain on the same day.) Think about what value you ought to select for b, within the constraints we've already established. If you choose a value other than one, then all I have to do is bet on this occurring and I am guaranteed a win. You will pay me $(1 - b)S$, which will be a positive sum. And I will never have to pay you back, because it's not possible for the event not to occur. But if you were to choose a value of one, on the other hand, then you wouldn't have to pay me anything if I bet on the event occurring. So you would be protected. Thus, we can conclude that b should be equal to one when b concerns (what we might call) 'a certain event'. (And if b concerns a statement, it should be equal to one when the statement is certainly true.)

Although it might seem surprising, we have already shown that b should satisfy two of the axioms of probability. (That is, on the axioms presented in Appendix A.) The first says that any probability must lie in the range of zero to one, and the second says that the probability for certain events or statements must be one. We could go on to derive the remaining axioms of probability, in a similar fashion, although we would need to consider a series of bets in order to do this. (If you're curious about the full details, you can have a look at the discussion in Gillies 2000: 59–65.)

2 Problems with the Dutch Book Argument

On the face of it, this Dutch Book argument – that degrees of belief should obey the axioms of probability – is good. It starts by asking us to consider situations with which we are familiar, namely betting scenarios, and ends with an unexpected result. But you may have some nagging doubts. We already noted that the betting scenarios might be difficult to implement, for example.

We should not be *too* picky about implementation, because idealizations are often necessary to bring theory into contact with practice. This is true in physics, as in philosophy and the social sciences. There are frictionless surfaces, molecules with no volume, weightless bodies, and so on. However, such idealizations are normally explicit in physics, in a way they may not be in the Dutch Book argument above. So let's begin by thinking a little more carefully about the hidden assumptions it rests on.

How about the stipulation that you (the bettor) do not have any information about whether I (the other betting party) will bet one way or the other? Does it follow that you would give an unbiased betting quotient b? You could have a really strong hunch that I will bet one way rather than the other, despite not having any evidence to that effect. And that hunch could lead you to select a value for *b* that is not, in your view, fair. I sometimes get hunches like this when playing Texas hold 'em poker, and choose to bet in cases where I would expect to lose judging purely on the visible cards and number of players remaining in the hand.

But might this not be considered irrational? Shouldn't you remain neutral about whether an event will occur, and assign each possibility (i.e. occurrence and non-occurrence) a degree of belief (or probability) of one half, if you have no information about whether it will occur or not? The worry is that an advocate of the *subjective* view of probability would not want to say this. It looks like an application of the principle of indifference, which we criticized during our discussion of the logical interpretation, in the last chapter.

An advocate of the subjective view would want to say, instead, that your degrees of belief about how I will bet are rational as long as they satisfy the axioms of probability. In other words, you will be rational provided that one is the sum of (a) your degree of belief that I will bet for the event occurring and (b) your degree of belief that I will bet against the event occurring, when you assume that (c) I will definitely bet one way or the other. (Well, almost. Several advocates of the subjective view also think that no one should have a degree of belief of one in a claim that is not a tautology, i.e. true by definition and/or the laws of logic, or a degree of belief of zero in a claim that is not a contradiction, i.e. false by definition and/or the laws of logic. This seems sensible; it is unwise to preclude possibilities compatible with one's information *altogether*, and similarly unwise to entertain impossibilities. But adding in this extra requirement, i.e. that (a) and (b) should be neither zero nor one, does not help in this case.)

So how about trying to fix the scenario by instead requiring that you must think I am just as likely to bet one way as the other? Unfortunately, this does not appear to work either. It makes it doubtful that we can measure degrees of belief reliably. After all, the way for me to determine if you *genuinely* think I am as likely to bet one way as another would presumably be to ask you to bet on it. So then we would be in another betting scenario. And I would want to make sure you think that I am as likely to bet one way as another in *that* scenario. So then I would need another test, with yet another gambling scenario. And so on, forever. Clearly this is hopeless.

The most natural way to avoid this problem, which is just to ask you how confident you are that I'll bet one way or another, is no good either. You have every reason to lie, because what you say may affect how I will bet. (We previously mentioned that you should not have any control over whether the event we are betting about will occur. By a similar line of reasoning, you should not have control, or influence, over how I will bet.) And even if you did not intend to lie, you might still not be aware of the correct answer. Why? As Ramsey points out, how strongly you feel about something need not indicate how strongly you believe it:

> [S]uppose that the degree of a belief is something perceptible by its owner; for instance that beliefs differ in the intensity of a feeling by which they are accompanied, which might be called a belief-feeling or feeling of conviction, and that by the degree of belief we mean the intensity of this feeling. This view would be very inconvenient, for it is not easy to ascribe numbers to the intensities of feelings; but apart from this it seems to me observably false, for the beliefs which we hold most strongly are often accompanied by practically no feeling at all; no one feels strongly about things he takes for granted. (1926: 169)

A natural reply is that we *must* know that we believe some things by introspection – by feelings – alone. But Ramsey does not deny this. He denies only that we can determine the *degree* of those beliefs by introspection. He proposes instead that: 'in many cases...our judgement about the strength of our belief is really about how we should act in hypothetical circumstances' (1926: 171). But this is not to say that these judgements are normally correct. If they were, then gambling scenarios would not be needed.

We will return to the issue of the measurement of degrees of belief shortly, in the next section. But beforehand, let's note that there are several further criticisms of the Dutch Book argument based on your (the bettor's) thoughts about how I (the other betting party) will behave. For example, maybe it is worth leaving yourself open to being Dutch Booked if you don't expect that I will exploit the opportunity to Dutch Book you? You could think I was incompetent, i.e. unlikely to spot your mistake. And maybe you stand to win more if you allow the *possibility* of being Dutch Booked, in the event that this possibility is not exploited.

Alternatively, you might think that my opinion about whether the event is likely to occur will be based on different information from yours (because you have special information to which I am unlikely to be privy). This point is subtly different from the one we considered above, about not having any information about which way I will bet. Think of the following situation. You do not have any information concerning how I intend to bet, but you are supremely confident that I don't know what you do about whether the event will occur. In fact, you are extremely confident that it will occur, on the basis of information you know that I do not have. Use your imagination. You know that a race will be fixed, and that a particular horse will win. But you also know that only you, and your close (tight-lipped) friends who are fixing the race, have this information. Should you assign a betting quotient of (approximately) one to that horse winning? Doing so will protect you from any losses in the event that I happen to bet on that horse winning. But there's a real chance, you think, that I will bet against the horse winning, *provided you do not declare that you are so confident*. (If you do declare such high confidence, it may lead me to *suspect* you know that the race is fixed, or something similar.) And maybe you think that's a risk worth taking. Maybe you want the chance to win some money. So while it is true that someone who was very *risk-averse* would choose a betting quotient of one on this 'certain event', to protect herself against loss, another rational person, who was less risk-averse, would not (under appropriate circumstances).

We can pull all the above criticisms together by noting that they have a common cause. This is the *gaming* aspect of two player (or more) gambling scenarios. In short, the problem for the Dutch Book argument, of the form presented above, is that playing a game well involves strategy. You must consider what your opponents will do. And you may wish to mislead your opponents about what you really think, in the hope of influencing what they will do to your advantage.

3 Measurement and 'Degrees of Belief'

But does being able to *measure* degrees of belief really matter? Doesn't the idea of a degree of belief make intuitive sense? Moreover, didn't some of the previous criticisms of using gambling to measure degrees of belief *rely* on the idea that there really are degrees of belief? For example, didn't we suggest that you might not disclose your degree of belief about some event happening *because of* other degrees of belief, e.g. about whether doing so would be a good strategy?

These are reasonable thoughts. However, they have some negative consequences for the Dutch Book argument. Even if it showed that your betting quotients should obey the axioms of probability, these would be disconnected from the degrees of belief that the subjective interpretation of probability concerns. So it would not indicate the importance of having one's degrees of belief obey the axioms of probability.

But perhaps this recognition suggests a radically different approach. What if we *forget* about degrees of belief in any mental sense? What if we just understand degrees of belief *to be* betting quotients? In fact, De Finetti advocates this idea, which we might call a betting interpretation of degrees of belief, in some of his early work. A simple version of the betting interpretation is that we should think of degrees of belief as actual betting quotients, given in actual betting scenarios.

We will see what's wrong with this view in a moment. Beforehand, however, we need to understand that there are deeper reasons, independent of the Dutch Book argument, for which De Finetti thought it was crucial for degrees of belief to be measurable:

> In order to give an effective meaning to a notion and not merely an appearance of such in a metaphysical-verbalistic sense an operational definition is required. By this we mean

> a definition based on a criterion which allows us to measure it. (1990: 76)

In this passage, De Finetti advocates *operationalism*. This was popular in the early part of the twentieth century, particularly among those more scientifically inclined. It is easy to see why. The underlying idea is that a precise understanding, in terms of measurements in the everyday world of action and experience, should be available for the concepts we deploy. Else, how can we really claim to understand those concepts? In the words of the physicist Percy Bridgman, 'we mean by any concept nothing more than a set of operations; the concept is synonymous with the corresponding set of operations' (1927: 5).

Unfortunately, this is a wrongheaded doctrine. Consider some of the simplest concepts we have, like 'one' and 'red'. Start with 'one'. Can you think of the set of operations with which this is synonymous? If you can't, as I can't, then presumably you don't explicitly understand the concept. But maybe you still implicitly understand it? In that case, though, you do not seem to need an explicit definition. (If you think carefully, you will realize that you can't define most of the words you use.) And it is not plausible that you (and I) don't understand what 'one' is, or only have a partial understanding of it.

Maybe you think that we haven't tried hard enough to define 'one'? Here's a standard operation in which it is important (and by which children perhaps learn the concept): counting. But it is hard to come up with a neat definition in terms of counting, without using the idea of number, or other numbers in particular. We could say that 'one' is always counted before 'two' when counting upwards, for example. But that relies on 'two', which is another number. So how about saying that 'one' says something about the tokens of a type you have before you when it is the terminal number you speak when performing a counting process of the tokens of that type before you? This seems too vague. What it says, after all, is something about the *number* of tokens. Hence, it does not seem that we have a handle on exactly what operations are relevant here. (Indeed, offering definitions in philosophy is, in general, an extremely tricky process.)

Perhaps picking on 'one' is a bit unfair, because an operationalist might think that there are distinctly mathematical operations, and that mathematics is a system of definitions not based on *physical* operations. So let's have a think about 'red' instead. What operations will we define it in terms of? Looking at things, perhaps? This is a good start, but really we need some kind of comparison operation. Ultimately, this is because to check whether something is red, I need to compare it to another red thing (or my memory of another red thing). But there are shades of red, of course; and we judge the similarity of shades to see if they count as the same colour. So what counts as the definitive 'red thing', on which our operations can be based, if anything? It seems crazy to suggest that there is such a single thing, as a matter of fact. So what are the operations? Here we have one of the simplest concepts we possess, and we are at somewhat of a loss about how to make it explicit. It also seems vague. Think of looking along a spectrum of colours, to the boundary between red and orange. There will be a point at which you are not sure whether the shade counts as one or the other. So even if there were some definitive 'red thing', consulting it would not always help.

We have seen that operationalism is a highly problematic doctrine, even if it was introduced for laudable reasons. Let's now see how it plays out in the context of degrees of belief. Recall De Finetti's idea that degrees of belief are closely related to betting quotients, and the simple version of this view, namely that degrees of belief are *actual* betting quotients. Now think of a person who hates gambling with a passion – maybe her mother was an avid gambler, who lost all of her family's money at the gaming table – and refuses ever to bet. Does she have no degrees of belief? We are forced to answer in the affirmative. But this violates our intuitive understanding of what a degree of belief is, in so far as this is based on our everyday understanding of what a belief is.

In fact, rather surprisingly, De Finetti admits that there is some justice to this complaint when he writes:

> The criterion, the operative part of the definition which enables us to measure it, consists in this case of testing,

> through the decisions of an individual (which are observable), his opinions (previsions, probabilities), which are not directly observable. (1990: 76)

The problem now is that 'opinions', which is associated with 'probabilities', seems not to be defined operationally. So there is no ultimate benefit to understanding degrees of belief as actual betting quotients, rather than as opinions *measured by* actual betting quotients.

Anyway, there is a sense in which some opinions, at least, are observable. I know about some of mine, just as you know about some of yours, by introspection. For example, I know I think that gay marriage is morally unobjectionable. You know whether you think the same thing or not. As we saw in the last section, however, this does not mean that we know the *strength* of our opinions by introspection. As Ramsey put it:

> [W]hen we seek to know what is the difference between believing more firmly and believing less firmly, we can no longer regard it as consisting in having more or less of certain observable feelings; at least I personally cannot recognise any such feelings. The difference seems to me to lie in how far we should act on these beliefs . . . (1926: 170)

Hence, the basic idea of Ramsey, instead, is that one's degrees of belief consist in the betting quotients that one is *disposed* to give, whether or not one ever has to bet. Let's go back to gay marriage to illustrate the point. I might realize by introspection that I find it morally unobjectionable, but be unsure (or wrong) about how strongly I believe that. I might not know exactly how firmly I would stand up for gay rights, in this respect, when challenged. (So although Ramsey thought 'the degree of a belief is just like a time interval; it has no precise meaning unless we specify more exactly how it is to be measured' [1926: 167], he did not support operationalism.)

But this illustrates a key concern about moving from actual to dispositional betting quotients. How I respond when challenged will depend not only on my opinions about gay marriage, but also on my other opinions and other of my personal attitudes. What if I think the person I'm talking to about gay marriage is a homophobic thug, who will thrash me if I disagree with him too strongly? And what if I have an intense fear of being thrashed? (You could also think back to the example of the person who hates gambling. Do we really want to say that such a person has *dispositions* to bet? At best, she only has hypothetical ones, i.e. ways she would bet *if* she did not hate gambling.)

Admittedly, this is an extreme example. And our natural response is to think that if I agreed with the homophobic thug, then it would only be in order to placate him. So I would just be *lying* about my opinion. But even in less extreme examples, a similar kind of worry can exist. The basic idea is that trying to measure someone's degree of belief can result in changing it. But before we come on to this, let's consider if we can measure degrees of belief without using gambling scenarios at all. After all, we've seen that these are especially tricky cases to deal with.

4 An Alternative to Gambling Scenarios for Measuring Degree of Belief: Scoring Rules

Worries about the gaming aspect of betting scenarios eventually led De Finetti to propose a new way of measuring degrees of belief (or for him, as we've seen, defining 'degrees of belief'). Ramsey passed away at the meagre age of 26, so had little time to change his mind. But there is every reason to think he would have done so, given his impressive contributions at such a young age. (I encourage you to look Ramsey up, and learn more about him. He was a fascinating character, who was the best performing mathematics student – the 'senior wrangler' – in his year group at Cambridge. Incidentally, the great Keynes, whom we encountered earlier, was only the twelfth best performing student in his year.)

Anyway, what's the different way of measuring degrees of belief? The fundamental idea is to use (what might be thought of as) a one player prediction game, with fixed scoring rules, in place of a gambling scenario. If you make accurate predictions, you'll be rewarded. If you make bad predictions, you'll be penalized. There is therefore an incentive to make accurate predictions, and to avoid bad predictions. Moreover, it may be argued that you should choose betting quotients in this *one player game* such that they satisfy the axioms of probability, lest bad predictions ensue.

Let's consider an example, to illustrate the idea. Imagine we told a meteorologist that she would be paid on the basis of such a scoring rule. Good predictions will lead to higher pay. Poor predictions will lead to lower pay. The onus is now on her to make her very best predictions. That is, with some relatively minor caveats. She should be satisfied that whether her predictions come true will be measured reliably, that the range in possible pay is significant for her, and so on. It is also easy to see how selecting betting quotients that did not obey the axioms of probability could be disastrous for her. If she predicted something that could not occur for logical reasons, e.g. that it would rain and not rain in the same place on the same day, then she would be *guaranteed* to lose money.

Now we could still, admittedly, have some doubts about this proposal. Imagine our meteorologist makes bad predictions, but then notices that one of her colleagues is repeatedly making much better predictions. Wouldn't she just copy that colleague's predictions, if she could? Yes. But then it might be argued that learning of her colleague's predictions would *change her own degrees of belief* about the future weather. So a defender of the scoring rule approach to measuring degrees of belief could say that she would be reporting *her* degrees of belief, but only after trying to make them more accurate. Such a defender could add that using scoring rules is not just a way of measuring degrees of belief. It is, moreover, a way to encourage people to try to make their degrees of belief 'better'. (For the moment, just note that making one's personal probabilities 'better' could be understood as meaning 'closer to the actual world-based probabilities' for a pluralist about probability. We will come to this in the next chapter.)

Besides, there are some situations in which people don't have an opportunity to copy others, or to look to others for advice. So the scoring rule approach to measuring degrees of belief appears to be superior to the (two player) gambling alternative. To be completely fair, though, we should admit that gambling scenarios *sometimes* measure degrees of belief accurately. The difficulty is just in working out when they do, and when they don't.

In summary, a balanced view on the measurement issue is as follows. Strengths of opinions (or degrees of belief) exist, although we find it hard to characterize them precisely, and can be measured reasonably accurately in the correct circumstances. How good a measurement we get depends on the instrument we use. And so we see how Ramsey's analogy between degrees of belief and time intervals is apt, at least up to a point. No one doubts that there is a fact of the matter about how long something takes (in a particular frame of reference). But we can measure that more or less accurately. One highly imprecise way is just to count 'one, two,...' at gaps thought to correspond to seconds. A better way is to use a normal analogue wristwatch. A better way still is to use a digital stopwatch. And so on, right up to the best available atomic clocks. But there is always a limit of precision.

The place where the analogy falls down is that we don't normally take measuring some interval to affect how long it actually is. But lots of physical measurements do involve disturbing the system that they target. Think about using a simple mercury thermometer to measure the temperature of a small volume of water. In order for the mercury to expand, heat has to be transferred from the water. So when we read the thermometer, we are not getting a reading of the temperature of the water at the moment we inserted it. In principle, the difference could be significant. In practice, e.g. when cooking, it normally isn't.

5 Objections to the Subjective View of Probability

We have now looked in considerable depth at the foundations of the subjective interpretation, and explored a key argument in its favour, namely the Dutch Book argument, as well as a promising (scoring rule) alternative. Let's now switch to a dialogue format, to explore some objections to the subjective interpretation.

STUDENT ONE: Here's something I find a bit weird. Do we really have degrees of confidence – or belief, whatever – so precise as to be written as numbers?

DARRELL: Good objection. Could you give an example to illustrate what you mean?

STUDENT ONE: Sure. Let's imagine I am choosing a betting quotient that will be closely linked to my genuine degree of belief – you know, my real opinion about whether something will happen. It could be on whether a horse will win a race. Are we really expected to believe that there is a precise numerical value – e.g. 0.51327834 – for that?

DARRELL: So you're getting at the idea that probabilities can take on infinitely many values, mathematically speaking, whereas our degrees of belief are more 'coarse grained'?

STUDENT ONE: Yes, 'coarse grained' is a nice way to put it. I guess 'quantized' is another. Just like energy differences in quantum mechanics, differences in degree of belief come in minimum amounts...

DARRELL: OK. Let's think about the consequences. It will be helpful, I think, if we now imagine a woman being offered a bet at some particular odds, and trying to decide whether she thinks the bet is fair or not...

STUDENT TWO: I see. What you're going to suggest is that we can imagine two bookies offering highly similar but not identical odds, on the same bet...

STUDENT ONE: ...and whether she – the bettor, I mean – would definitely think one bet was fair and the other wasn't?

DARRELL: I see that I am not really needed today! This kind of thought experiment is exactly what you need. Why don't you go ahead and explain what it tells us?

STUDENT ONE: I think that the difference in odds could be insignificant for her. The bets could be offered at odds of a million to one and a million and one to one, for instance. She might think both bets were *fair*, even if she would prefer one set of odds rather than another if she were to bet.

STUDENT TWO: I agree. But there would also be a point at which the difference would be too big for them both to count as fair, right? Maybe it would be a million to one as against nine hundred thousand to one, or something like that?

STUDENT ONE: Yes. So we could understand her degree of belief as an interval.

STUDENT TWO: Makes sense.

STUDENT ONE: But now I wonder if the boundary – you know, between bets that are clearly fair and bets that clearly aren't fair – is always sharp. Actually, we can think about just one bet being offered to see this. The odds could be chosen such that the bettor is not sure whether they are fair or not.

DARRELL: OK. Good. So we may have to tinker with things a bit, to refine the subjective interpretation. But there's no serious problem here. Any more objections?

STUDENT THREE: Yes. I like the logical interpretation we looked at before, even if there are some problems with using the principle of indifference. Why? It makes a serious attempt to capture the idea that there is an objective fact of the matter about whether an argument is good even when it is not valid. Now...

DARRELL: Sorry to interrupt, but I think I know where you're going. I just want to set the scene a bit. Stop me if I have the wrong end of the stick.

STUDENT THREE: Ha! Philosophers seem to love to interrupt. OK. Go ahead.

DARRELL: Right. Let's begin by noting that many people describe some non-deductive arguments, like inductive arguments, in terms of probability. In place of validity, that's to say, there's another notion where high probability of the conclusion, given the premises, makes the argument 'good'.

STUDENT THREE: I quite agree.

DARRELL: So let's assume that if p entails q, then $P(q, p)$ should be one, for any person, even on a subjectivist view. Well, roughly. It could be a bit more complicated; for example, some might add that the person would have to recognise that p entails q, for it to follow that $P(q, p)$ should be one... But let's put the detail on this to one side. The situation when p does not entail q is different.

STUDENT THREE: Yes! Imagine q is 'Darrell is a philosopher' and p is 'There is life on Mars'. No sane person would think these two propositions are connected in any way. But on the subjective view, a *rational* person could think $P(p, q)$ is almost one. So they would think the fact you're a philosopher is strong evidence for life on Mars. That is, in the absence of any other relevant assumptions.

STUDENT TWO: That's nuts. So what counts as evidence, except in deductive cases, is totally subjective?

STUDENT THREE: Yup.

DARRELL: It is a standard objection, and a strong one. But let me try to put it in context. Shouldn't our degrees of belief satisfy the axioms of probability? That is, even if you think there are *some* situations in which there are extra rationality requirements?

STUDENT THREE: Yes.

DARRELL: Ok. So aren't you really saying the subjective interpretation is fine as far as it goes? Isn't your objection just that it doesn't go far enough?

STUDENT THREE: I suppose so.

DARRELL: Well as we will see shortly, this can be fixed by adding further rationality constraints on degrees of belief... That's to say, we may require that degrees of belief do more than satisfy the mathematical axioms of probability, in order to count as rational.

STUDENT ONE: May I add something else?

DARRELL: Sure.

STUDENT ONE: I find the subjective view pretty attractive. Just because lots of people think that many of the same non-deductive arguments are good, this doesn't mean there is anything deeper to it. There may be explanations for why they agree about this kind of thing, cultural or even evolutionary, without good inductive arguments existing in any objective sense. The subjective interpretation is good for explaining what actually happens, say in science. Scientists respond to what they *take* to be good evidence, and look for the same.

STUDENT THREE: OK. But wouldn't we like to think science was on a surer foundation than that?

STUDENT ONE: Maybe we would *like* to. But that doesn't make it so!

STUDENT TWO: Here's a better argument. Wouldn't it be a miracle for science to be so successful if people were just acting on unfounded opinions when it came to (a lot of) evidence?

DARRELL: You have opened a can of worms! We can't hope to answer this complicated question now. But the way our discussion has gone shows how important our understanding of probability is for what we think about how, and why, science works. We will return to this topic in the final chapter.

So, to summarize, the main worry about the subjective interpretation is that it allows for too wide a range of *rational* difference of opinion. Or, to put it differently, it seems to make it too easy to have rational degrees of belief. As mentioned in the dialogue, we will come back to this issue shortly. You need only wait until the next chapter.

6 Subjective Monism and Independence

We have almost finished our coverage of the subjective interpretation. But before we move on to look at the next alternative, it is worth considering whether the subjective interpretation can stand alone, as the one true interpretation of probability. De Finetti thought it could, and spent a lot of time trying to show how. But was he successful?

To understand De Finetti's attempt fully, we would need to get technical. But to keep things simple, I will instead give you a flavour of what he tried to do, and the main difficulty he encountered. Let's start with the following question. 'What is the best objection to the view that all probability is subjective?'

The answer is that there are patterns of events that are accurately predicted by the use of mathematical probabilities. Take casinos as a case in point. Why are they so successful? Why aren't they regularly forced to close down, by customers repeatedly winning large sums of money? Why are such customers incredibly rare?

To see the point more clearly, let's consider the game of roulette, which originated in France. This is played on a wheel, with several numbered and coloured compartments of equal size, as shown in figure 4.1. The wheel is spun. A ball is rolled on a track around the rim of the wheel in the opposite direction (e.g. anticlockwise if the wheel is spinning

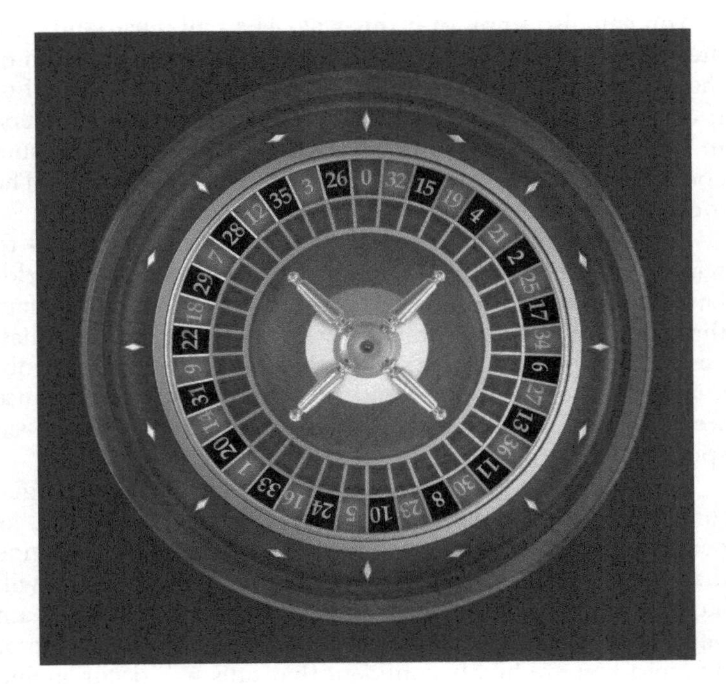

FIGURE 4.1 A European roulette wheel

clockwise). The winning bets are determined by which compartment the ball lands in.

Let's begin by considering what happens if you bet that the ball will land on a specific number, such as seventeen. Casinos offer odds of 35:1, which would be fair if the probability of landing on a particular number were 1/36. But there are thirty-seven numbers. So if the (world-based) probability of landing on each number is equal, fair odds are actually 36:1. (The zero is there for a reason. And roulette wheels in the USA have a double zero compartment, in addition, to stack the odds even more in the house's favour.)

Another way to illustrate the casino's advantage is to imagine that you place a bet for the same amount on each individual number. One of your bets will win. But it will not cancel out the losses you incur. If you bet $1 on each number, for example, you would receive $36 for a total stake of $37.

You can also think of it this way. The ball must land on a number (if the bet is to be completed). Therefore, the sum of the probabilities of all the possible results should be one. But if a probability of 1/36 is assigned to each of the numbers, in line with the odds offered by the casino, then the sum comes to *over* one. Clearly, something fishy is going on. The 'odds' offered violate the axioms of probability.

But what does De Finetti say about why casinos tend to win by offering bets with such odds? He argues that world-based probabilities are merely an illusion caused by the fact that many people have similar (or identical) degrees of belief. So although it may appear to us that roulette involves probabilities in the world, independently of us, this is just because we have similar confidence in particular results on a given spin, or over a number of spins.

De Finetti does not stop there, though. He goes on to argue that widespread agreement is to be expected over time, as people encounter more evidence. To illustrate, let's imagine that we disagree strongly, to begin with, about what will happen when a brand new coin is flipped. Let's say I am highly confident that heads will occur, giving P(H) a value of 0.9, and you are highly confident that tails will occur, giving P(H) a value of 0.1. De Finetti argues that as the coin is flipped and more information comes in, our subjective probabilities – our rational degrees of belief – will come closer together. So if heads happened roughly half the time, over the course of many flips, we would eventually meet in the middle and each have P(H) approximately equal to 0.5.

This result can be guaranteed if we must both learn – that is to say, change our opinions – by the same mechanism in order to be rational. And for De Finetti, part of this mechanism is Bayesian updating. (The full mechanism is Bayesian updating *with exchangeable initial probabilities*; I will explain exchangeability shortly.) In line with my policy of keeping it simple, I am not going to produce Bayes's theorem, on which it is based. (But I do explain Bayes's theorem in Appendix B, with a worked example, if you want to take a look.) It's easy to get De Finetti's basic idea. All *rational* people change their degrees of beliefs in the same way, as follows. They begin with an idea about how probable some event (or other hypothesis) is, independently of any direct experience concerning it. But as new evidence comes in, they become interested in new conditional probabilities. Return to the previous example of the brand new coin, to see the idea. (And reread the discussion of conditional probabilities in the previous chapter, if you find this hard to follow.) Initially, when asked whether the coin will land on heads, I am interested in P(*H*, *b*), where *b* represents my background information. This is known as the *prior probability* of *H*. Later, when I have some data on previous flips of the coin, I will be interested in P(*H*, *b* & *e*), where *e* represents the data on the previous flips. This is known as the *posterior probability* of *H*.

Now Bayes's theorem gives a neat way of working out P(*H*, *b* & *e*) on the basis of P(*H*, *b*) and one other value, namely P(*e*, *H* & *b*). This is known as the *likelihood* of *H* and *b* on *e*. Think of it this way. If the probability of heads occurring were 0.9, then one would not expect the first twenty flips of the brand new coin to result in tails. So if the result were tails on the first twenty flips (*e*), then one would reassess the probability of heads occurring (assuming one held one's other background information, *b*, constant).

However, even if we all learn in a Bayesian way, it does not follow that we will all gradually come to agree on our probability assessments, as our shared evidence increases. De Finetti requires also that our initial probability assignments should be *exchangeable*.

Imagine I am considering how five flips of a coin might turn out. My probability assignments are exchangeable if I think that the probability of *n* heads (or equivalently, 5 – *n* tails) is the same no matter how that happens. Let's consider one head, which can happen in five different ways: HTTTT, THTTT, TTHTT, TTTHT, and TTTTH. If De Finetti were right, I should think that P(HTTTT) = P(THTTT) = P(TTH TT) = P(TTTHT) = P(TTTTH). And so on. Each permutation should be given an equal probability. (We discussed permutations and combinations back in Chapter 3, in the example of the coin flipping game that I played against the lawyer.)

The problem with this assumption, however, is that it is too restrictive. To see this, imagine you have a box in front of you with two lights, red and green, which is connected to a computer. You know that one of the lights will flash every few seconds, but not which one. Now imagine that, as you watch, you see the following (where R represents red and G represents green):

R G R G R G R G R G R G R G R G R G R G R G R

Most people would conclude that there was a pattern. In fact, it is plausible that the (world-based) probability of a green result next is one, because the computer is programmed to flash one light and then the other, continually. But if we are stuck with assuming exchangeability, we could only come to conclude, after many trials, that the probability of a green result after a red result is a half. That's because the green light shows half of the time.

So in effect, requiring exchangeability for one's personal probability assignments looks like (blindly) assuming independent world-based probabilities exist in the case under consideration. Two events (or propositions) are independent, recall, if whether one occurs (or whether one is true) does not affect whether the other occurs. Coin flips are like this. But traffic lights, which inspired the example above, are not. Indeed, there are many examples of dependence in the real world. For example, the probability of snow in England on a day picked at random is lower than the probability of snow in England on a day *after* another on which snow fell.

So De Finetti did not manage to produce a convincing argument that all probabilities are subjective. And it is unclear why we should want to think they are, anyway. The only obvious advantage would be that we could interpret all probability statements in the same way (without stopping to think about how they were being used). This would be a matter of convenience at best.

Further Reading

An excellent intermediate to advanced level text on subjective probability is Jeffrey (2004). An intermediate level overview is provided by Gillies (2000: ch. 4). Ramsey (1926) is accessible (at an intermediate level), and worth revisiting; this is reprinted, along with other useful advanced level discussions – e.g. by Kyburg – and editorial analyses, in Eagle (2011). Eriksson and Hájek (2007) provide an intermediate to advanced level discussion of degrees of belief.

5
The Objective Bayesian Interpretation

In Chapter 3, we looked at the logical interpretation. This has the attractive feature that each (numerical) probability has a unique value, corresponding to a relationship between propositions (or sets of propositions). Its main unattractive feature, however, is that it's unclear how we are supposed to get at those values.

Then, in Chapter 4, we looked at the subjective interpretation. This has the attractive feature that it is clear – at least, much clearer – about how we can measure probabilities. Its key unattractive feature, however, is that probabilities do not typically have unique values. So there is no objective probability of relativity theory given the scientific data at our disposal. There is no fact of the matter, independently of us. There are only several personal probabilities about relativity. And that seems counterintuitive.

So what about taking the good bits from each, and forging a new interpretation of probability? This is what objective Bayesians, like Edwin Jaynes (1957) and Jon Williamson (2010), aim to do. They use the subjective interpretation as a starting point, thereby linking probabilities to measurable degrees of belief. But they introduce additional requirements for measurable degrees of belief to count as probabilities. To put it simply, objective Bayesians and subjectivists *agree* that probabilities are rational degrees of belief. But they *disagree* about what it takes for a degree of belief to be rational. Objective Bayesians think it requires obeying several more rules than subjectivists do.

1 Extra Constraints on Degrees of Belief

Consider a group of degrees of belief. Let's say they concern rolls of a regular tetrahedral die. (I use this example because four-sided dice are not as common as six-sided dice – although they are used in a number of tabletop role-playing games – and because it is important that you do not assume any experience of rolling dice, other than what I specify, in what follows.)

In figure 5.1, we can see that the die has landed on 4. But how can we determine the rational degrees of belief concerning how this die will land, when rolled? Williamson (2010) says that these should obey three constraints:

1 *Probability*
2 *Calibration*
3 *Equivocation*

Let's take these in turn. *Probability* says that the rational degrees of belief should satisfy the axioms of probability. And this is no different from what a subjectivist would say. If we assume the die must land on one of its four faces when rolled, for example, then the probability of it landing on any one of the four must be one: $P(1) + P(2) + P(3) + P(4) = 1$. Similarly, the probability of the die *not* landing on any one of the four faces must be zero: $P(\neg 1 \,\&\, \neg 2 \,\&\, \neg 3 \,\&\, \neg 4) = 0$. And so on. No surprises here.

Calibration is a new addition. It says that the rational degrees of belief should also be sensitive to any other relevant information gained. In particular, they should be sensitive to observed frequencies of (pertinent) events, i.e. evidence about world-based probabilities. So imagine the rational degrees of beliefs are conditional on the information that a '1' result has occurred 40 per cent of the time on all the rolls of the die so far. P(1) should be set accordingly: $P(1) = 0.4$. Other available

FIGURE 5.1 A tetrahedral die
[slpix]

empirical information, and physical theories tested using that information, might also be considered relevant, although we are *not* including that here. (For example, evidence concerning the results of rolls of other regular convex polyhedral dice – such as the standard cubic variety, and the others depicted in figure 5.2 – could also be taken into account. Considerations of physical symmetry might suggest that all such dice are fair [when they are of uniform density, i.e. not loaded]. So we might think the probability of landing on any given side will be $1/n$, where n is the number of sides of the regular convex polyhedron in question.)

The final constraint, *equivocation*, says, in the words of Jaynes, that we should be 'maximally noncommittal with regard to missing information' (1957: 623). The idea should be familiar from the discussion of the principle of indifference, in Chapter 3; as Keynes put it, recall, 'equal probabili-

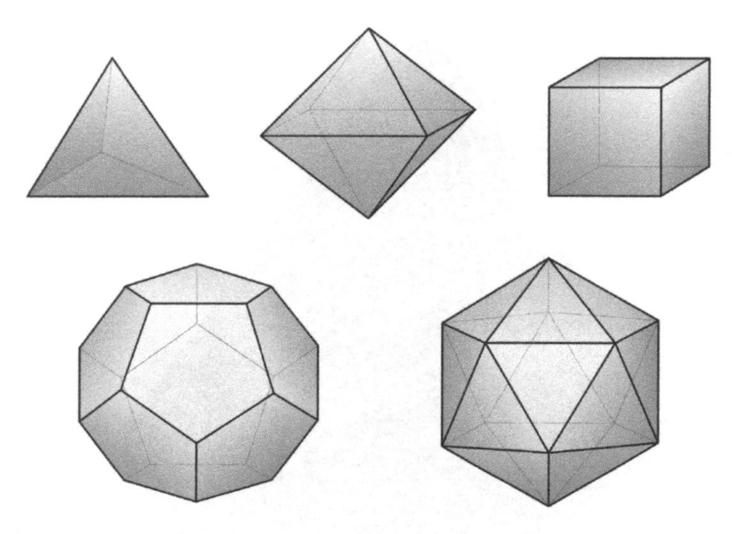

FIGURE 5.2 The Platonic solids
[Peter Hermes Furian]

ties must be assigned to each of several arguments, if there is an absence of positive ground for assigning unequal ones' (1921: 42). Let's stick with the example of the regular tetrahedral die, to illustrate. P(1) = 0.4 on the empirical information assumed when *calibrating*, but there is no information about which of the remaining possibilities will occur. So a rational person should *equivocate* over those remaining possibilities, i.e. assign each an equal probability; P(2) = P(3) = P(4) = 0.2.

2 Objective Bayesianism in Action: A Further Illustration

Before we continue, it may help to have another example of how the constraints above can be used to calculate probabilities. Imagine you know the following two claims are true, from experience: $p \vee r$ and $q \oplus r$. (We used '⊕', which stands for 'or' in an exclusive sense, back in Chapter 3. The new symbol, '∨', stands for 'or' in the *inclusive* sense. The differ-

TABLE 5.1 Truth table for $p \vee r$, $q \oplus r$, and $p \oplus (q \oplus r)$

p	q	r	$p \vee r$	$q \oplus r$	$p \oplus (q \oplus r)$
T	T	T	T	F	T
T	T	F	T	T	F
T	F	T	T	T	F
T	F	F	T	F	T
F	T	T	T	F	F
F	T	F	F	T	T
F	F	T	T	T	T
F	F	F	F	F	F

ence between the two senses is simple: $p \oplus q$ is true only when one of p or q is true, and the other is false; $p \vee q$ is false only when p and q are both false.) What rational degree of belief should you have in – i.e. what probability should you attach to – the claim that $p \oplus (q \oplus r)$?

Table 5.1, which shows how the truth-values of the formulae interrelate, will help to answer. Remember first, from our use of a similar truth table in Chapter 3, that the rows represent logical possibilities. So the eight rows, taken together, represent all the possible combined values of p, q, and r (and the other formulae). But only three of these rows are relevant to answering the question, because you know that $p \vee r$ and $q \oplus r$ are both true. These are rows two, three, and seven; they are marked in grey.

Think of it this way. You want to know if $p \oplus (q \oplus r)$ is true. And the information you have – that $p \vee r$ and $q \oplus r$ are true – helps you to *rule out* some possibilities. It tells you that the actual world is not represented by rows one, four, five, six, or eight. You can think of this as a logical position-finding process, and the truth table as a map. Your information is like a rudimentary GPS device. It lets you know, in this case, that you are in one of three logical places. If you did not have the information, you would only know you were in one of eight logical places.

You are now calibrated. Beyond this, however, you have no information about which logical place you are in. So you must equivocate over the remaining possibilities (while respecting the probability constraint). Hence, you should assign each remaining possibility an equal probability; P(row two) = P(row three) = P(row seven) = 1/3. Now you need only note that $p \oplus (q \oplus r)$ is true only if row seven corresponds to the place you're in. (In rows two and three, it's false.) Hence $P(p \oplus (q \oplus r), (p \vee r) \& (q \oplus r))$ – the rational degree of belief you should have in $p \oplus (q \oplus r)$, given $p \vee r$ and $q \oplus r$ – is one-third.

A potential argument for the equivocation norm also becomes apparent from this example. Assume which of the three logical positions you find yourself in – those consistent with your empirical information (shown in grey in table 5.1) – is selected *at random*. Now, over time, if you repeatedly found yourself in the same kind of situation, $p \oplus (q \oplus r)$ would be true one-third of the time (and false in the remainder). So considering frequencies over the three logical possibilities – rather than frequencies of events or things in the real world – is the trick. A key worry about this argument is whether the starting assumption – that you find yourself in some position *at random* – is correct. (After all, there's no *process* by which you were dropped into one of three logically possible situations. No demon is playing a game with you. Hopefully!)

3 Is Objective Bayesianism an Interpretation of Probability?

In short order, we'll look at some criticisms of objective Bayesianism. And we'll also explore its relationship to the logical interpretation, which we haven't yet said much about. But beforehand, we should cover a potential point of confusion. Let's use a dialogue.

STUDENT ONE: Wait a minute. I'm confused.

DARRELL: About what?

STUDENT ONE: Aren't we supposed to be discussing *an interpretation of probability*?

DARRELL: Yes.

STUDENT ONE: But objective Bayesians appear to be interested in numbers – rational degrees of belief – that do more than satisfy the mathematical axioms of probability.

They have to satisfy calibration and equivocation norms too.

DARRELL: Good point. But on the one hand, many groups of numbers satisfy the axioms of probability theory without us wanting to say they are probabilities. And on the other, Keynes even went so far as to suggest, although I did not mention this earlier, that some probability relations are non-numerical.

STUDENT ONE: I just had the impression that we were in the business of interpreting the mathematical notion, in operation...

DARRELL: I did set things up that way, in Chapter 1, didn't I? But you can look at the situation as follows. The objective Bayesian view is designed to treat how the mathematical notion should be used in *many* contexts.

STUDENT TWO: Let me get this straight. An objective Bayesian could admit that some *mathematical* probabilities should be interpreted as just subjective? For example, if an agent had degrees of belief that satisfied the probability constraint but not the other two?

DARRELL: Yes, sure.

STUDENT ONE: OK. I still wonder if thinking of things in terms of 'interpretation' is a bit misleading, though.

DARRELL: That's not unreasonable. And we definitely don't want to get into a verbal squabble about how to use the word 'probability'! So don't think in terms of interpretations, if it suits you.

STUDENT ONE: How about thinking of matters like this? The argument between subjectivists and objective Bayesians is about the *role* of the mathematical theory of probability in explicating how people should reason.

DARRELL: Sounds good to me. Either way, we're doing philosophy of probability.

In summary, one option is to think of the mathematical notion of probability as a means to describe something else, such as an idea of probability that we possessed before the mathematics was worked out. Alternatively, if you want to reserve 'probability' for the mathematical notion, then you can see us as exploring legitimate ways to apply the notion.

4 Objections to Objective Bayesianism

We can now move on to examine some objections to objective Bayesianism. Let's stick with the dialogue format.

STUDENT ONE: I have a worry about the calibration norm.

DARRELL: Yes?

STUDENT ONE: It seems strange to say that a person strongly opposed to empiricism – who therefore believes that observed frequencies are not relevant to forming judgements about what will happen in the future – is irrational if he doesn't pay attention to observed frequencies. Do you see what I'm getting at?

DARRELL: Interesting point. It's like failing to pay attention to frequencies is irrational! And that's a hard thing to believe. However, I think objective Bayesianism – at least in some variants – *can* allow people to have, and act on, evidence that frequencies should be ignored in certain situations.

STUDENT TWO: Maybe we can also look at 'irrational' from an external perspective? We can think of 'being irrational' as reasoning in a way that is not reliable, rather than not internally consistent, or something like that...

STUDENT ONE: I suppose so. But now I wonder if subjectivists and objective Bayesians ultimately understand 'rational' in different ways, and even take deeply different epistemological viewpoints.

DARRELL: That's an insightful suggestion. I believe some of their differences of opinion can, indeed, be explained in that way.

STUDENT TWO: Interesting. Is it OK if we move on?

STUDENT ONE: Yes.

STUDENT TWO: My worry is slightly different, but connected. It's about the relative priority of the calibration and equivocation norms.

DARRELL: That sounds impressive!

STUDENT TWO: Ha! I hope so! Anyway, here's the idea. Remember when we talked about dependent events, at the end of the last chapter? You know, with the alternating red and green lights? Well, the calibration norm seems to suggest we should *assume* that events are independent unless we have evidence that they are dependent.

DARRELL: Can you give an example?

STUDENT TWO: Sure. Consider rational degrees of belief on the outcomes of rolls of a six-sided die, when you know only that a '3' result has happened 50 per cent of the time previously. Why should you think the probability of a '3' result on the next roll is 0.5?

DARRELL: What you are saying is pretty subtle. You have to agree that we do have some relevant information – that is, at least, if we assume that the future will resemble the past in some respects – about how the die will land?

STUDENT TWO: Yes, but my worry is whether I should act on the information *with respect to the result of the next roll*. Why can't I reason as follows? I have no information about whether the results of the rolls are dependent or independent. So according to the equivocation norm, I should give dependence and independence equal degrees of belief, of one half...

DARRELL: Your point is well made. Objective Bayesians say one should equivocate only after calibrating. But it looks like what's involved in calibration – when it comes to observed frequencies – goes against the spirit of equivocation, and therefore undermines the justification for it. Right?

STUDENT TWO: Right. Keynes's logical view seemed much more flexible on this issue.

DARRELL: Yes. But let me add this much. In principle, an objective Bayesian could just leave 'calibration' much more vague, and not explicitly mention *anything* about frequencies – although that would, I think, result in a different variant of objective Bayesianism from the one we've been considering.

STUDENT THREE: Hmm. And now equivocation and Keynes have been mentioned, can I raise a further worry?

STUDENT TWO: Shoot.

STUDENT THREE: Does objective Bayesianism *really* do any better than the logical interpretation? First, isn't the equivocation 'norm' really just the principle of indifference, in another guise? Second, you said 'calibration' might be left vague. Isn't this just what Keynes does in his version of the logical view?

DARRELL: This is the objection I was waiting for.

This raises the question of whether objective Bayesianism differs from the logical interpretation as much as it may first appear. We will cover this in the next section.

5 Objective Bayesianism vs. the Logical Interpretation

STUDENT ONE: You think that the objective Bayesian view isn't much different from the logical view. So why did you devote a chapter to it?

DARRELL: I see you've read one of my articles! You're right. I don't think there is much difference. But I devoted a chapter to objective Bayesianism for two reasons. First, I might be wrong! Second, 'objective Bayesianism' is the name typically associated with an active position in the philosophy of probability; few people, by comparison, claim to be working on a 'logical' view.

STUDENT TWO: OK. But why do you think the two positions aren't much different?

DARRELL: Let's argue about it, shall we? But allow me to begin by admitting that they are different in one genuine respect. The logical interpretation ultimately concerns objective logical relations – of partial entailment or content – between propositions. Objective Bayesianism directly concerns (rational) degrees of belief, instead. But if you think that rational degrees of belief should respect logical relations (like Keynes), or that logical relations should be *defined* in terms of rational degrees of belief (like Williamson), then the connection becomes clear.

STUDENT ONE: I see. But before we continue – is it defensible to hold a logical interpretation of probability and *deny* that rational degrees of belief should respect logical relations?

DARRELL: Yes. Consider the work of Popper. He argued that the logical probability of universal scientific laws is zero relative to any evidence we can get. The basic idea is that any finite evidence is compatible with infinitely many theories; so if we equivocate over those theories, we'll have to give them each a probability of zero. But does this mean that it is irrational to believe in universal scientific laws, like those of thermodynamics? Not necessarily. Maybe it's OK to believe in something without evidence, provided there's no evidence against it. How about for pragmatic reasons? You see the idea, I hope.

STUDENT ONE: Yes, I do. I think of Pascal's wager: that it can be useful to believe in God, because it may help you to gain entry to heaven, even if you don't have any evidence that God exists. And what you've said is compatible with a kind of 'halfway house' view, where rational degrees of belief should *sometimes* respect the logical relations, but other times should – or need – not.

DARRELL: Spot on.

STUDENT TWO: So let's get back to why you *don't* think the logical and objective Bayesian views are different. Or why you don't think there are any other *significant* differences, to be more accurate.

DARRELL: Why don't you tell me how you think they *are* different?

STUDENT TWO: OK. The logical interpretation doesn't involve the calibration norm. How about that?

DARRELL: Actually, this is not so clear as it may first appear. Why? It is possible to understand the calibration norm as a *rule* about logical relations between propositions (which might contain empirical information). Go back to your example in our last discussion, about the six-sided die. We know only that '3' has occurred half the time on previous rolls, and that there are six possible results. (OK, we know a bit of maths too. Use your common sense.) But maybe the logical relation between these propositions and 'Three will be the result of the next roll' is 0.5? And maybe calibration is just a general rule that covers similar cases?

STUDENT TWO: I see your point. So you could believe in calibration and have a logical view?

DARRELL: I think so. You could argue it's not required for the logical view – that adding calibration leads to a particular *version* of the logical interpretation. But there's no reason to reject the logical interpretation because you think calibration is appropriate.

STUDENT TWO: That makes it clear. The logical view doesn't imply calibration is false, so you can't show the logical view is false by showing calibration is true.

DARRELL: Nicely put. Who's next?

STUDENT THREE: I was going to say that there's no clear way to measure degrees of belief in the logical view of Keynes. But now I think your reply will be that he doesn't rule out measuring degrees of belief by gambling procedures, or scoring rules, or what have you. So really, Keynes could just embrace whatever view he wanted about measurement of degrees of belief – that is, as opposed to logical relations.

DARRELL: Quite right.

STUDENT THREE: Then maybe we can go back to the issue of whether the equivocation norm is something different from the principle of indifference?

DARRELL: Yes, let's do that. In fact, objective Bayesians use a principle that they call 'the principle of maximum entropy'. This was proposed by Jaynes.

STUDENT THREE: OK. So what is the statement of the principle of maximum entropy?

DARRELL: Well entropy is an idea from physics – Jaynes was a physicist – and I don't want to get too technical. So let me instead just quote him, when he describes how he thinks the principle of maximum entropy is different from the principle of indifference:

> The principle of maximum entropy may be regarded as an extension of the principle of insufficient reason [another name for 'the principle of indifference'] (to which it reduces in case no information is given except enumeration of the possibilities...), with the following essential difference. The maximum-entropy distribution may be asserted for the positive reason that it is uniquely determined as the one which is maximally noncommittal with regard to missing information, instead of the negative one that there was no reason to think otherwise. (Jaynes 1957: 623)

STUDENT THREE: That's pretty wordy.

DARRELL: Yes, so let's take it in bits. Jaynes makes two key claims: (A) the principle of maximum entropy reduces to the principle of indifference when only possible outcomes are known; and (B) there is a justification for applying the principle of maximum entropy – namely, that it results in maximally noncommittal degrees of belief – that there is not for applying the principle of indifference.

STUDENT ONE: Can we tackle (B) first? I know we are supposed to be arguing against you, but it seems wrong to me. I am looking at the definition of the principle of indifference given by Keynes, which you quoted in Chapter 3. Here it is:

> The Principle of Indifference asserts that if there is no known reason for predicating of our subject one rather than another of several alternatives, then relatively to such knowledge the assertions of each of these alternatives have an equal probability. Thus equal probabilities must be assigned to each of several arguments, if there is an absence of positive ground for assigning unequal ones. (Keynes 1921: 42)

DARRELL: Thanks for raising that.

STUDENT ONE: You're welcome. Now look at the last sentence: 'equal probabilities must be assigned to each of several arguments, if there is an absence of positive ground for assigning unequal ones.' Clearly, Keynes and Jaynes agree about this, right?

DARRELL: Right!

STUDENT ONE: And what's more, Keynes's statement of the principle does not say we should assign equal probabilities *because* there is no known reason for doing otherwise. It says we should assign equal probabilities in the absence of known reasons for doing otherwise.

DARRELL: Right again. There is no 'because' in the statement!

STUDENT ONE: So Jaynes is wrong that the principle of indifference is asserted for a negative reason. (B) is false. In fact, one argument for using the principle of indifference is *exactly* that it avoids presuming something is true (or false) when there's no information about whether it is.

DARRELL: I suppose I should now ask you what you think about (A), since you're doing my job for me so well.

STUDENT ONE: It's false too.

DARRELL: Why?

STUDENT ONE: It's dead simple. You can have, in Keynes's words, 'absence of positive ground for assigning unequal' probabilities when you have more than an enumeration of possibilities.

DARRELL: An example would be?

STUDENT ONE: Easy. I have seen a coin flipped many times, and know the ratio of 'heads' to 'tails' results is approximately 1:1. Now I know more than just the possible outcomes of the next flip. But grounds for assigning unequal probabilities to 'heads' and 'tails' are still absent. So the principle of indifference makes a positive recommendation to assign equal probabilities.

DARRELL: I agree.

STUDENT ONE: So Jaynes is way off? How come?

DARRELL: Let me be brutally honest. I don't think Jaynes read Keynes, to whom he nevertheless refers, carefully enough. So he didn't give credit where it was due, or appreciate that he was reinventing the wheel to some extent. That's my opinion. Others might disagree.

STUDENT THREE: And that's what philosophy is all about!

DARRELL: You took the words right out of my mouth.

STUDENT TWO: Well, I still think you're being a bit harsh.

DARRELL: Maybe. But even if we accept that the principle of indifference and the maximum entropy principle are different, what would stop Keynes, or any other fan of the logical view, from embracing the latter? Keynes *certainly* does not deny that there is a positive reason for equivocating!

STUDENT TWO: OK.

STUDENT THREE: Sorry to interrupt. But can I now redirect our attention to the issue I raised before, in objecting to objective Bayesianism? If there's no real difference between the recommendations of the principles, then surely they both encounter paradoxes like the 'horizon' one covered in Chapter 3? That's to say, if we carve up the possibilities differently then we will get different probability assignments by applying the principles?

DARRELL: I think so. Actually, Jaynes devoted a lot of time to trying to tackle such paradoxes. And I don't think he was successful. Williamson, on the other hand, just accepts that some of them are insoluble. He acknowledges that equivocating can lead to more than one result in *some* situations. But he still thinks we should equivocate.

STUDENT THREE: OK, thanks for explaining that.

DARRELL: No problem. The upshot, of course, is that probabilities don't *always* have unique values in Williamson's version of objective Bayesianism. In the 'horizon' case, for example, there is no unique answer one has to have in order to be rational, *because* one can legitimately equivocate in different ways.

STUDENT THREE: I see. So such paradoxes are more problematic for the logical view than they are for the objective Bayesian one – or Williamson's version thereof, at least?

DARRELL: Perhaps. But maybe it is possible to have a logical view while denying that a logical probability obtains in such cases? Or to insist that it's incredibly hard to calculate, or something like that.

STUDENT THREE: That's food for thought.

6 From Subjectivism to Objective Bayesianism: A Spectrum

As may have become apparent in the dialogue above, there is really an interpretative *spectrum* between the subjective view and the objective Bayesian view. Starting with subjectivism and the probability constraint on rational degrees of belief, one is free, in principle, to introduce whatever extra constraints one likes. One could introduce the calibration norm without introducing the equivocation norm, or vice versa, for example. And one could take a more sophisticated approach, such as stipulating context-specific norms. Here's an example. You could say that equivocation is only a norm when faced with finite indivisible options – as in the gambling scenario in the pub, where I won so much money, that we covered in Chapter 3. Or you could say something subtly different, like that equivocation is only a norm when you're faced with options you *believe* to be finite and indivisible. The possibilities are endless. The choice is yours.

Further Reading

Key advanced level works articulating and defending objective Bayesianism are Jaynes (2003) and Williamson (2010); the latter is considerably more accessible, and is intermediate level in parts. Childers (2013: ch. 6) provides a detailed intermediate level discussion of the maximum entropy principle. Rowbottom (2008) considers the relationship between the logical and the objective Bayesian interpretations at an intermediate to advanced level.

6
Group Level Interpretations

Almost all the philosophers we've so far encountered agree on the following. Individuals have beliefs. And they have degrees of belief – or degrees of confidence in their beliefs – too. Rational people's beliefs and degrees of belief obey a set of rules, and probability theory plays a part in specifying those rules.

But can't groups have beliefs too? It appears so. The use of 'We believe' is natural. (Google it!) It occurs regularly in statements produced by groups, such as scientific research teams and political parties. So should we not also expect groups to have degrees of belief? 'We strongly believe', 'We are highly confident', and 'We are certain' are natural phrases too. And it is natural to think that such phrases are used to express (high) degrees of *group* confidence in claims.

Thus we are left to suspect that groups are also susceptible to being irrational in a similar way to individuals, if their degrees of belief don't obey particular rules. And this prompts us to consider whether there are group level interpretations of probability, just as there are personal level interpretations (such as subjectivism and objective Bayesianism).

The first such group level interpretation, the intersubjective interpretation, was proposed by Donald Gillies (1991). Let's start off by looking at the motivation behind this. We can then move on to consider some alternative group level views, which I have developed more recently (Rowbottom 2013b).

1 Group Dutch Books

Imagine you are a bookie. Your aim is to make a combination of bets that ensures you will profit no matter what happens. And an ideal way to do this is to find different customers who will accept different betting odds on the same event. To make matters simple, let's consider two bets. You could think as follows:

> Customer A will pay R, in return for S if E occurs.
> Customer B will pay U, in return for T if E does not occur.
>
> If E occurs, you will gain U and lose $S–R$.
> If E does not occur, you will gain R and lose $T–U$.
>
> Therefore,
> If U is greater than $S–R$ and R is greater than $T–U$, you will profit whether or not E occurs.

Let's consider a concrete example. Imagine you get customer A to pay you $50 in return for $60 if E occurs, and customer B to pay you $50 in return for $60 if E does not occur. Whatever happens, you will keep $40. In effect, you will profit from the failure of A and B to accept the same betting quotients concerning E. To see this, recall (from Chapter 4) that R may be broken down into bS, where b is the betting quotient on E in the bet accepted by customer A. So U can also be broken down into $(1–b^*)T$, where b^* is the betting quotient on E in the bet accepted by customer B.

A and B have been Dutch Booked *as a group*, in the example above. And this is because b is not equal to b^*. The betting quotients do not satisfy axioms of probability. The probability of E and the probability of not-E should sum to one, but $b + (1–b^*)$ is greater than one.

2 Group Dutch Books and Rationality

What is the significance of this kind of group Dutch Book, when it comes to questions about rationality, and rational degrees of belief in particular? If A were the same person as B, in the example above, he would have made a foolish mistake. That is, provided he had made the two bets without a change in his information about whether E would occur. (If he discovered new evidence to make him highly confident that E would not occur, after making the first bet that it would, then he might want to make the second bet to try to recoup some of his expected losses. He would expect to lose $40 when E did not occur, rather than $50.)

If A and B are different people, however, then they must be appropriately connected in order for their betting behaviour to count as irrational. Imagine you're A, and I'm B. Why should you care if the bookie makes money whatever happens? All you care about is that *you* make a profit, by winning *your* bet against the bookie. It is irrelevant that for you to win, I have to lose (because I happen to have bet differently from you).

But now imagine that A and B are Romeo and Juliet, a married couple with shared income and resources. They have a joint bank account, from which the money for each of the bets comes. So the end result of the two bets is bad for both of them. They lose $40 from their joint account, which would have been better spent on dinner and a movie together.

From the point of view of either Romeo or Juliet, the behaviour may be understandable. Maybe neither knew about the other's bet, or expected the other to bet. But might they still have shown poor decision-making skills *as a couple*? It seems so. Imagine they had mobile phones to hand, and could easily have consulted each other about their bets, in order to coordinate them. They could easily have avoided a sure loss. They could also have discussed whether E would occur, and pooled their information on the issue. They had open communication channels. In summary, wise couples discuss investment choices, with the aim of reaching an agreement about what to do, *when it is easy to do so and they are investing joint funds*. They definitely try to avoid making investments, as a partnership, that result in sure losses.

We have seen that (at least) two conditions are required for considerations of group rationality to be present, when bets by different people are considered: (a) there must be shared funds involved in the bets, and (b) there must be a way for the bettors to communicate before making the bets. This is a rough version of what Gillies (1991) says. We could demand a bit more precision in several respects. We could ask whether information must be able to flow both ways, for example. (I will leave this as an exercise for you. Think about this, for starters: even if A could not communicate with B, A might still be able to send B instructions not to bet in a way that would conflict with his own bet.) But the basic idea is clear.

Two points are worth emphasizing, however. First, 'funds' need not be interpreted literally, as money. It is not even necessary to think in terms of material goods, like houses or cars. The 'funds' could be anything shared by the people, which they both value (and therefore have an interest in); and the shares need not be equal. Economists often use the concept of utility, to capture this. The utility of an hour's hill running might be greater for me than it is for you, whereas the utility of eating a hamburger might be greater for you than it is for me. So you might prefer the latter to the former, if given a choice, whereas I might choose the other way. (I'm a vegetarian and enjoy hill running. But your preferences are *probably* different.)

Second, 'bets' need not be interpreted literally either. This idea should be familiar from Chapter 4. As Ramsey (1926) insisted, we are making bets, in a sense, all the time. When I chose to take a temporary job at Oxford University, and gave up the opportunity to be interviewed for several permanent jobs elsewhere, I was 'betting' that I would find the experience extremely rewarding, and that it wouldn't damage my career in the long run. And when you choose what to eat for dinner this evening, you will similarly 'bet' that you'll enjoy it more than the other alternatives available.

A final example will help to illustrate the previous two points. Imagine three partners in crime, who learn that the police will soon be visiting their homes to arrest them. They get together, and form a careful plan about what to do. They agree to tell similar stories about their involvement in the relevant incident, when questioned by the police, in order to avoid being charged with the crime. (They recognize that telling *exactly* the same story would be suspicious. The police know not to expect testimony to be exactly the same.) Each partner has a reason to act as agreed, namely to avoid the disutility of being charged, and perhaps being convicted. And this is true even if the partners in crime would be liable for different penalties upon conviction. Perhaps one of the partners in crime has already been convicted of two serious crimes, and would suffer from a severe 'three strikes and you're out' penalty if convicted again. Maybe the other partners have never been convicted of committing a crime before, and would be treated more leniently by the judge. But it is still in their interests to cooperate. Each member of the group stands to lose – via fines, community service, or time 'inside' – if they do not.

3 The Intersubjective View: Gillies on Group Degrees of Belief and Consensus

We have seen that there are situations where it is important for groups to reach a consensus, in order to avoid being Dutch Booked. Or more precisely, we might say: 'A group is protected against Dutch Books by using betting quotients that satisfy the axioms of probability.' We think a group is rational if it protects itself against external exploitation, and irrational if it fails to protect itself, provided it has a reasonable opportunity to do so.

But let's now ask how this relates to the idea of a *group degree of belief*, with which we started. We noted that use of 'We believe' is natural. However, this does not tell us whether groups *really* have beliefs, above and beyond the beliefs of their members, or how so-called 'group beliefs' relate to individual beliefs. Do we really need to talk about group beliefs at all? Or can we rest content with talk of group betting quotients?

Gillies writes:

> [M]any, if not most, of our beliefs are social in character. They are held in common by nearly all members of a social group, and a particular individual usually acquires them through social interactions with this group...It is actually quite difficult for individuals to resist accepting the dominant beliefs of a group of which they form a part, though of course dissidents

> and heretics do occur... [A]s well as the specific beliefs of a particular individual, there are the consensus beliefs of social groups. Indeed the latter may be more fundamental than the former. (2000: 169–70)

Think about the importance of testimony, in order to appreciate Gillies's point. How many of the things you know, or think that you know, did you learn independently of other people? Not many, and certainly not most. You were schooled formally, and by your parents (or other appropriate parties). You also learnt from friends and acquaintances. And even now, you are using my testimony in order to form new beliefs. No doubt you have thereby accepted several of the beliefs of the community of philosophers of probability, of which I am a part.

But why does Gillies say that social beliefs 'may be more fundamental' than individual ones? He does not think that social beliefs are *ontologically independent* of individual beliefs. Rather, social beliefs depend, for their existence, on individual beliefs: a social belief that *p* exists only in virtue of the existence of many individual beliefs that *p* (in members of the society). Instead, Gillies thinks that social beliefs are resistant to change in a way that (merely) individual beliefs are not. Thinking of matters causally may help. If almost everyone believes that *p*, in a community, this will tend to cause a new member of the community to come to believe that *p* (over time). It is highly unlikely that a new member who happens to believe not-*p* will succeed in persuading the group that not-*p*, i.e. in changing the group belief. (Note that the effect of one group's beliefs on those of another group is different, and is partly a matter of scale. Think, for example, of the influence that the scientific *community* has on public beliefs. Still, strong resistance can be present when the claims of scientists conflict with those of the public. Think of how the claim that the Earth revolves around the Sun was once seen as incorrect, because it conflicted with the biblical story.)

By extension, we can say that group degrees of belief are identical to shared individual degrees of belief. And this fits with what Gillies says of intersubjective probabilities: 'Intersubjective [Interpretation]: Here probabilities represent the degree[s] of belief of a social group which has reached a consensus' (2000: 179). Consider any proposition, *p*, and any group, G. G does not have a degree of belief in *p* unless the members of G have reached consensus on *p*. And G does not have a probability on *p* unless G has a degree of belief on *p*. Therefore, G does not have a probability on *p* unless the members of G have reached consensus on *p*. This is Gillies's view.

And what does consensus on *p*, in G, require? On Gillies's view, each and every member of the group must share *exactly the same individual degree of belief* that *p*. Otherwise, *on the assumption that betting quotients correspond to individual degrees of belief*, the members of the group might be willing to accept individual bets on *p* that would leave the group open to a Dutch Book.

Consider the following example, based on one of Gillies's (1991: 529–30) own, to understand how we can use intersubjective probabilities *understood as* consensus degrees of belief that satisfy the axioms of probability. Imagine there are two competing research groups, G_1 and G_2, working in the same area of physics. There is also a dissident scientist, D, who works alone in the same area. The members of G_1 are working on showing that their favourite theory, T_1, is correct. The members of G_2 are instead trying to show that T_2, their preferred alternative, is better. Finally, D believes that neither T_1 nor T_2 is a good theory, and is out to refute both.

The situation could be as follows. All the scientists – those forming G_1 and G_2, along with D – might (rationally) agree that T_1 and T_2 are just as good at predicting the existing evidence, *E*. So $P(E, T_1) = P(E, T_2)$, when the probabilities are intersubjective and concern *all* the scientists working in the field. (Consider the following possible theories in the most fundamental and important area of science, namely rabbit science: T_1 is 'All rabbits are black or white', whereas T_2 is 'All rabbits are black or brown', and *E* is 'The 1000 rabbits observed so far are all black'.) However, the different parties may still disagree about how plausible those theories are independently of the evidence. $P(T_1) > P(T_2)$ on the intersubjective probability of G_1, whereas $P(T_1) < P(T_2)$ on the intersubjective probability of G_2. And D thinks $P(T_1) = P(T_2) = r$, but also that r is very low; this is D's *subjective* probability. So for D, $P(\neg T_1 \,\&\, \neg T_2) >> P(T_1 \vee T_2)$, although G_1 and G_2 share the opposing view that $P(\neg T_1 \,\&\, \neg T_2) << P(T_1 \vee T_2)$. Or in plain English, for those not familiar with logic, G_1 and G_2 doubt that both T_1 and T_2 are wrong, whereas D is confident that both T_1 and T_2 are wrong. (Note that I used '$\vee$' above, for 'or' in the inclusive sense, because it is possible that both T_1 and T_2 are true, on the evidence E. All rabbits could be black.)

This quick example indicates how intersubjective probabilities – understood as consensus degrees of belief – may be useful when thinking about how science works, or indeed about other situations in which groups and individuals interact, e.g. in politics or business. But I shall argue below that intersubjective probabilities can, and should, be thought of in a broader way.

4 My Alternative View: Consensus on *Using* Betting Quotients

I agree with Gillies that there are intersubjective probabilities. But I do not think that they necessarily reflect group degrees of belief, or even group beliefs. So let me begin by offering a positive argument for my alternative view, which is that intersubjective probabilities should be understood as consensus *betting quotients*. I will use a thought experiment.

A general and his leading officers meet to discuss a battle plan. They have common interests in winning the battle, and in minimizing their army's losses. The general presents the information available on the enemy troops and position, as well as the information available on his own forces. He then asks the officers for their opinions about the appropriate strategy to adopt, and the appropriate tactics to use in a variety of possible scenarios that might occur as a result. The discussion is long, and heated. There is little agreement between the officers about the best way to proceed, even after arguments are presented. Time is up. The enemy is on the move. The general has to make a decision. And so he does. On the basis of what he thinks to be the best arguments, he issues orders. He outlines the plan. His leading officers accept those orders, and agree to implement the plan.

As the officers are leaving the meeting, to return to their units, they vent their opinions on the plan. Some think it acceptable, but not ideal. Others think it reckless. All the officers agree on is that the plan is consistent, and that their units will be working towards the same goal. Nevertheless, they all solemnly swear that they will implement the plan, because they know that if each instead did their own thing, the result would be a disastrous loss. In the common interest, that is to say, they agree to work as a team.

In this case, the group has agreed to use common betting quotients in those contexts in which group interests are relevant, and in which the group is vulnerable to exploitation from outside forces. And if those betting quotients violated the axioms of probability, the group would suffer. This could happen, for example, if the plan was for an artillery unit to target a specific location with fire, and for friendly infantry to occupy the same location beforehand. (This is effectively a bet that only enemy forces will be in a location, and a bet that allied forces will be in a location. That is, with some unwritten assumptions accepted: that it is undesirable to kill one's own troops, that artillery fire on an area will kill many of the troops in the area, etc.)

The officers do not share the same beliefs, let alone degrees of belief, about how the battle will unfold (and what is best to do as a result). They disagree, for example, about how the enemy army will deploy. But they will all act *as if* the enemy will deploy in the way the general thinks. (So in effect, they are agreeing to act on the general's relevant personal degrees of belief, although they have had the opportunity to influence these.)

Moreover, the officers *do not even share the same personal betting quotients* on how the battle will unfold (and therefore on the wisdom of the plan dictated). It is easy to see this. Imagine now that some of the officers remain for a while, after the meeting proper. They continue arguing amongst themselves, although they have accepted the general's verdict. Overhearing them, the general grins enigmatically, and offers each a bet. If the battle is won, the bettor will be promoted. But if the battle is lost, the bettor will be demoted. Some officers eagerly take the bet. Some are not sure about whether to take it or not. Others flatly refuse it. And this is not just because some of the officers are more risk-averse than others. Rather, it's because they have different expectations about how the battle will unfold, given the plan they have agreed to implement. (The general is being rather wily, too. He hopes that those who accept the bet will fight even harder, to make sure a win happens.)

I conclude that it is reasonable to use probabilities to represent (rational) group betting quotients, rather than (rational) group degrees of belief. And this is not to deny that there are *some* occasions on which group degrees of belief match group betting quotients.

5 Gillies vs. Rowbottom: A Dialogue

Let's now use a dialogue to consider the relative merits of Gillies's intersubjective view and my alternative. In doing so, the differences between these will become more apparent.

DARRELL: So does anyone want to defend Gillies's view? He was a teacher of mine, as it happens, so I think someone ought to!

STUDENT ONE: Sure. Actually, I think you have missed a more charitable way of interpreting his account of intersubjective probabilities.

DARRELL: What's that?

STUDENT ONE: Well let me begin by quoting a passage from his main article on this:

> If a group does in fact agree on a common betting quotient [which, in combination with other such common betting quotients, satisfies the axioms of probability], we shall call that betting quotient the *intersubjective* or *consensus* probability of the social group. (Gillies 1991: 517)

DARRELL: Well read. But you don't deny that Gillies talks a lot about group beliefs, and group degrees of belief?

STUDENT ONE: I don't deny it. But have you considered the possibility that Gillies *assumes the definition of* group degrees of belief as common betting quotients, and indeed *assumes the definition of* individual degrees of belief as personal betting quotients?

DARRELL: Yes, I have considered it. I am not sure if it's right, but I did notice the following passage at the end of the same article, which suggests that it might be:

> The real questions regarding subjective and intersubjective probabilities are these. Can a theory of human belief with a definition of degree of belief in terms of the Dutch Book argument become an important and successful theory of psychology (or perhaps sociology)? Can such a theory help to explicate the beliefs of scientists, and their judgements of confirmation? (Gillies 1991: 532)

STUDENT ONE: I spotted that too. I think it shows that Gillies understands degrees of belief in the operational way that I suggested – as betting quotients.

DARRELL: I am not so sure, because just above he cites the discussion of Ramsey on degrees of belief, which we covered in Chapter 4. Then, as you will have seen, he writes:

> X's beliefs affect X's behaviour in all sorts of ways, but to render a belief measurable, we have to select just one particular observable effect of this belief. The effect in question is the choice of betting quotient which X would make if forced to bet under the conditions specified earlier; and we use this effect as our measure. (Gillies 1991: 532)

STUDENT ONE: OK. So maybe Gillies's assumption is *really* that the betting scenarios he discusses *measure* degrees of belief effectively.

STUDENT TWO: Yes. And if you are arguing against that, Darrell, then aren't we revisiting our earlier critical discussion of the Dutch Book argument for the subjective view, in Chapter 4?

DARRELL: I see your point. But think of my view this way. I do not deny that betting scenarios *ever* measure degrees of belief effectively. I just hold that sometimes, even when they do not, rational group *betting quotients* should still obey the axioms of probability.

STUDENT TWO: And hence, there must be a legitimate interpretation of probabilities as rational group *betting quotients*?

DARRELL: Exactly.

STUDENT ONE: I see. But doesn't that mean that intersubjective probabilities on Gillies's view are a *special case* of intersubjective probabilities on your view?

DARRELL: Yes. And there is another important difference between our views that we haven't yet touched on...

STUDENT TWO: Which is?

DARRELL: I argue it can be rational for the same person to use different betting quotients, on the same bet, depending on whether she is betting with personal or group resources.

STUDENT ONE: That's a bit quick!

DARRELL: Go back to the example of the general and the officers. How the officers will bet depends on whether the bets just concern them – e.g. personal future promotion or demotion – or the army as a whole.

STUDENT ONE: Mind if I offer another, simpler, example? That is, to make it clearer?

DARRELL: Not at all! Please do so.

STUDENT ONE: OK. Let's imagine a minimal group, of two. They are twin brothers, aged 18. They have a trust fund, left to them by their late mother. They have to choose which stocks to invest in. They will each get half of the money in the fund when they reach 30 years of age.

STUDENT TWO: Nice! So I guess they disagree about which stocks to invest in?

STUDENT ONE: Yes! They disagree about which stocks to invest in, but they have to invest. One brother wants to invest in company A. The other wants to invest in company B. But companies A and B are bitter rivals in the same marketplace, and it's clear that if A succeeds, B will fail, and vice versa. So they shouldn't just invest half of the current money in each.

DARRELL: So what do they do?

STUDENT ONE: They agree, instead, to invest in a third company, C, which supplies products used by both A and B. Both think there is less money to be made this way, but see that compromise is the only way to protect their joint funds.

STUDENT TWO: And the story does not end there, I guess?

STUDENT ONE: No, it doesn't. Each twin also has personal savings, and wants to show the other that he was right about how they should have invested. One puts all his personal savings into company A. The other puts all his into company B. Neither thinks that company C is a superior investment option.

STUDENT TWO: And none of the behaviour in the story is irrational!

DARRELL: An excellent example! We can make it tighter by saying there are only three possible companies to invest in.

STUDENT TWO: We could also make the betting quotients explicit with some small alterations. Say investment in A or B will result in doubling the money invested, provided the company remains afloat, by the time the twins are 30. Then equal investments in A and B will result in no profit, or worse, on the assumption that one will remain afloat at most.

STUDENT ONE: Yes. And we could add that investment in C is expected to increase money invested by only a quarter, provided it remains afloat.

STUDENT TWO: Absolutely. So one twin thinks that the bet on A is the only fair bet. The other twin thinks that the bet on B is the only fair bet. Neither twin personally thinks that the bet on C is *fair* – because they don't think it will pay enough if C succeeds, given the risk – but they agree to make that bet in the interest of protecting their joint funds.

DARRELL: Very nice! Maybe I should use your examples in future.

STUDENT ONE: Perhaps you will pay us commission? But actually, I'm not finished. I just thought of something else...

DARRELL: Which is?

STUDENT ONE: On your view, it doesn't matter if the twins' individual degrees of belief – on the matter under consideration – satisfy the axioms of probability. They can have an intersubjective probability even if they don't.

DARRELL: That's right. So intersubjective probabilities can exist in the absence of subjective probabilities, on my view.

STUDENT TWO: This does seem like an advantage.

6 From Intersubjective to Interobjective Probabilities: Another Spectrum

In closing, let's consider one further way that the intersubjective interpretation of probability can be modified. For all we've said so far, group betting quotients (and/or group degrees of belief) count as group probabilities simply when they provide the group with immunity from being Dutch Booked. But this is consistent with the group's consensus on betting quotients (or degrees of belief) being caused in numerous different ways, many of which would be entirely irrational. What if the group were cult members, brainwashed in order to serve the cult leader? Or what if they used random numbers in order to generate betting quotients, and simply got lucky (in selecting random numbers that satisfy the probability calculus)? Must we say that group probabilities are present?

It is reasonable to answer in the negative, and to require that consensus *must be achieved in a particular way* in order for group probabilities to result. For example, one might require the group to pool relevant information on the subject matter on which betting quotients are to be assigned. One might also require that group members with relevant expertise (if any) take a lead in the decision-making process. And so on. In Rowbottom (2013b), I consider several such possibilities, but do not advocate any in particular. My aim is mainly to open our eyes to the options. Just as there is a spectrum of alternatives to the (strict) subjective interpretation, so there is a spectrum of alternatives to the (strict) intersubjective interpretation.

One might even think that *rational* group betting quotients (or degrees of belief) have unique values. (Perhaps, following objective Bayesians like Williamson, this is only in finite cases). There may be a single correct process (or set of equivalent processes) by which to decide on betting quotients, which results in a single value for the betting quotients (if performed correctly). Thus it is even possible to propose an *interobjective* interpretation of probability as an alternative to the (strict) intersubjective interpretation, just as there is an objective Bayesian alternative to the subjective view.

Further Reading

Group level interpretations of probability are relatively new, and not widely discussed. (Nevertheless, I believe I've shown why they're interesting and potentially important.) The key places to look for intermediate to advanced level treatments have already been mentioned above: Gillies (1991), Gillies (2000), and Rowbottom (2013b).

7
The Frequency Interpretation

At the beginning of the book, we considered the difference between information-based and world-based approaches. And since then, we have focused on the former. But we have seen that they suffer from some difficulties, particularly when it comes to explaining why some things happen, or tend to happen. Casinos make money, almost without fail. (I have invested in casinos in Macau, on the Hong Kong stock market, for this reason.) But how could we explain this in terms of information, e.g. in terms of the (rational) degrees of the beliefs of gamblers and casino operators? Surely whether people believe that casino games are fair is irrelevant to whether they *are* fair.

If you need convincing, imagine that casinos all over the world began to lose money to gamblers on a regular basis. Initially, we would put this down to 'luck'. But what if it continued to happen? We would next suspect the gamblers of cheating. And what if we then investigated, and determined that no systematic cheating was occurring? Would we keep our estimates of the probability of winning on any individual kind of game, such as blackjack, the same? No. We would adjust them.

As we saw previously, it is always open to a monist who favours an information-based view of probability to insist that we would change our probability estimates *only* because our information – about the results of games in casinos, etc. – would change. But an advocate of world-based probabilities may respond by altering the thought experiment. Imagine a casino where there are no believers. The games are automated. The gamblers are robots. Is there not a fact of the matter about whether the games are fair? Would there be a probability of winning in any of the games, even if there were no believers left in the universe? 'Yes' seems to be a reasonable answer. (Even now, there are computer programs that buy and sell stocks. So something like this hypothetical scenario could happen in the future. Imagine we will set up the casino, so that the robots can gamble on behalf of their owners, but will later be wiped out by a virulent plague.)

1 Finite Empirical Collectives and Actual Relative Frequencies

On the world-based view, probabilities are 'out there' in the physical world. They do not depend on us, or propositions or language, for their existence. And it seems to follow that we learn of them, and their values, only by experience. But what, exactly, should we understand these world-based probabilities to be?

All the answers we will consider in this chapter are based on the idea that probabilities concern groups of things, or *collectives*. As the scientist and mathematician Richard von Mises (1928: 18) put it: 'FIRST THE COLLECTIVE – THEN THE PROBABILITY.' And since we're dealing with *world-based* probabilities, it seems wise to begin by considering *world-based* groups of things, i.e. *empirical collectives*. Such collectives can involve 'mass phenomena' as well as repetitive events, i.e. cases where 'either the same event repeats itself again and again, or a great number of uniform elements are involved at the same time' (von Mises 1928: 12).

One such collective is that of the rabbits born to date. It is finite. Let's say it has n members. What are the probabilities involving this collective? One view is that this depends only on the *attributes* of the rabbits therein, such as colour and size, and the frequencies of these *attributes* within the collective. Take 'a rabbit is black' as a case in point. If there are m black bunnies, out of the n bunnies in the collective, then its probability is m/n. This is just the *actual relative frequency* of black bunnies (over all bunnies that have ever existed). Clearly, this value can be no less than zero, and no higher than one. And it is easy to see that $n - m$ rabbits will not be black, if m bunnies are black, so that the actual relative frequency of non-black bunnies will be $n - m/n$, which is equal to $1 - m/n$. Moreover, the actual relative frequency of black bunnies *or* non-black bunnies will be $(m + n - m)/n$, i.e. one. So these actual relative frequencies satisfy the first and third axioms of probability (discussed in Appendix A). Similar considerations allow us to show that the other axioms are satisfied. (Imagine we want to find the relative frequency of *furry* black rabbits. We can multiply the relative frequency of black rabbits by the relative frequency of furry rabbits *among those that are black*. Say half of the rabbits are black, and half of black rabbits are furry. It follows that a quarter of the rabbits in the collective are furry and black.)

It is important to note, at this point, that the probability of *a* rabbit being black, in this example, has nothing to do with the probability that *the next* rabbit born will be black. Such a rabbit is *not* a part of the collective we began with. You might think there is an easy way to fix this. We could instead consider the collective of all rabbits born until now *plus* the next rabbit to be born. True. But this would still not provide a probability that 'the next rabbit born will be black'. It would just mean that we were considering the *overall frequency of black rabbits in a bigger collective*. So on this view, there is no probability that the last rabbit born was black, or that *any* particular rabbit was born black. The probability concerns only the actual frequency of black rabbits in the finite empirical collective. If the reason for this is somewhat unclear, don't worry. We will return to this issue later.

Basing probabilities on finite empirical collectives has a clear advantage over some of the other interpretative strategies we've previously considered, such as the logical one. Measuring probabilities by observation becomes unproblematic in many everyday cases, and is always possible in principle, if not in practice. Go out. Watch. Record the frequency. Learn the probability, or (at least) get more data on what it is. It is easy to see why this perspective on probability might appeal to scientists.

Unfortunately, the view that probabilities are relative frequencies (of attributes) in finite empirical collectives seems to conflict with how we often think of probabilities, and with how we typically talk of probabilities. First, consider a coin that has never been, and will never be, flipped. Is there a probability that *it* lands on heads when flipped? It seems so. And *must* this probability be information-based? It seems not. It appears that there could be a truth about the probability, grounded somehow in the world, even if we do not ever learn anything about its value. (Note that we cannot just consider the collective of all coins. To generate a frequency for heads, say, we have to consider collectives of coin *flips*. So although we could consider all coin flips made, *with any coin*, this still would not involve the coin in the example above.)

Second, consider a coin that is only flipped a small number of times, and is then destroyed. Imagine it lands on heads every time. Should we accept that the world-based probability that this coin lands on heads, when it is flipped, is one? Again, this seems wrong. We are inclined to think that we could have learned more about the probability, by flipping the coin more often. So we appear to be interested in something other than the *actual* frequencies in finite empirical collectives.

Third, and finally, there are many other curious consequences of accepting that probabilities are frequencies of finite collectives. Alan Hájek (1997) provides a long list. Here are just two items from that list. Imagine we know that the probability of a specific die landing on six, when it is rolled, is one-sixth. We can infer that the number of times the die has been (or will be) rolled is a multiple of six. Moreover, amazingly, we can also work out that *any die must be unfair* if it is not (or not to be) rolled, in total, for a number of times that is a multiple of six. Indeed, *all dice are unfair*, when we consider the collective of all rolled dice, provided that the total number of dice rolls is not a multiple of six.

If this isn't incredible enough, note also that all probabilities, on this view, must be rational numbers. (A rational number is a number that can be expressed as n/m, where n

and m are both whole numbers, and m is not zero.) But quantum mechanics involves probabilities that are not rational. So one has to accept that quantum mechanics is false, or that the probabilities therein are not world-based, on the basis of this definition. All from the armchair.

2 Infinite Empirical Collectives and Actual Frequencies in the Limit

Several of the problems identified above can be avoided by requiring that world-based probabilities be associated, instead, with infinite collectives. Imagine a coin that lands only on heads, over infinitely many flips. The probability of a heads result, on flipping the coin, is intuitively one. We do not need to worry about what would happen on any additional flips.

So why not define probabilities in terms of relative frequencies in infinite collectives? Mathematically speaking, there is a useful limit operation that can be used. To see how this works, consider the following formula: $(x^2 - 1)/(x - 1)$. What is its value when x is one? There is no answer, because 0/0 is undefined (or indeterminate). But we can nonetheless consider the value of the formula as x gets nearer and nearer to one, as shown in Table 7.1.

It is clear that $(x^2 - 1)/(x - 1)$ *approaches* two as x *approaches* one. So we say that the *limit* of $(x^2 - 1)/(x - 1)$, as x approaches one, is two. And we now have the tool that we need to think about values at infinity. Consider $1/x$, as a

TABLE 7.1

x	$(x^2 - 1)/(x - 1)$
0.999	1.999
0.9999	1.9999
0.99999	1.99999
1.00001	2.00001
1.0001	2.0001
1.001	2.001

simple case in point. What is its value as x approaches the limit of infinity? As illustrated in Table 7.2, it's zero.

TABLE 7.2

x	$1/x$
1000	0.001
10000	0.0001
100000	0.00001
1000000	0.000001

Mathematically, then, there's no problem in talking about relative frequencies in the limit of infinity. But there is an obvious problem with defining probabilities in such a way, if we only consider *empirical* collectives. We can find *some* empirical collectives that seem to be genuinely infinite; think of the average velocities of Earth, over ever decreasing periods of time, for example. But we use probabilities in many cases where the collectives are not, and could never be, infinite. And we might not want to say that these are all information-based. Think back to the motivation for a world-based view given at the beginning of this chapter. We want to be able to talk about games of chance in casinos, and explain why – or, at the very least, predict that – these will result in the house accumulating a much higher percentage of wins than the gamblers over time. But such games of chance are finite in number. And they always will be, it would seem.

It is open to an advocate of this view of probability to argue that the games will carry on indefinitely. But the underlying problem is deeper. Do we really have to know *whether* such events will carry on indefinitely in order to know whether we can think of (and use) probabilities concerning them in a world-based way? It seems not. Such speculations do not enter into the minds of casino owners, insurance companies, and the like. They just rely on data concerning the actual frequencies in order to guide their actions.

A related worry is as follows. Why should the actual frequencies tell us *anything* about the frequencies in the limit? Think of coin flips. Why won't these all become *heads* after time t, e.g. tomorrow? What guarantees that our current data is relevant, if anything? Stay tuned – or, at least, read on – for the answer, which will appear in the next section.

There is one final problem. Some infinite sequences do *not* have limiting values! Hájek (2009: 220) gives the example of the following sequence of coin tosses:

HT HHTT HHHHTTTT HHHHHHHHTTTTTTTT...

The pattern repeats. Next there will be sixteen (2^5) heads and sixteen (2^5) tails. Then thirty-two (2^6) heads and thirty-two (2^6) tails. And so on. Hence the relative frequency of heads (or of tails) does not stabilize, as the number of flips approaches infinity. It fluctuates endlessly, up and down, as illustrated in Table 7.3.

TABLE 7.3

Number of flips	Rel. frequency of heads
6	1/2
10	7/10
14	1/2
22	15/22
30	1/2
46	31/46

The low point is 1/2 all the way through, because there are continually stages at which there are as many heads as tails, but never points at which there are more tails than heads. The oscillations *do* become slightly smaller, but the high point only approaches 2/3 as n approaches infinity. (You can see that 15/22 is closer to 2/3 than 7/10, and that 31/46 is closer still, if you use a calculator. Later values include 2047/3070 and 8191/12286.) So when we think about what happens at infinity, we can conclude only that the relative frequency of heads oscillates between 1/2 and 2/3.

So how are we to know that the empirical collectives we encounter do not involve sequences such as these, where relative frequencies fail to have limit values? There does not seem to be anything incoherent in imagining that a coin lands in the way detailed above. And it is easy to see that there are other such sequences, e.g. where there are 3^n heads followed by 3^n tails. (Indeed, there are infinitely many other such sequences. Just replace 3 by any other natural number you like.)

3 Hypothetical Frequentism and von Mises's Relative Frequency Interpretation

We began by thinking about empirical collectives, and considering whether probabilities are just *actual* relative frequencies in these. But we found this view to be problematic for both finite and infinite collectives. A key problem is that probabilities often appear to be absent, when they should be present. For example, a coin could not be fair, or even have a probability of landing on heads, if it were never flipped.

There is a simple way to solve this problem. It is to consider *counterfactual possibilities*, or possibilities that are contrary to fact. Indeed, we do this regularly in our daily lives. We say things like 'If I had worked harder at school, I would have got better grades!' and 'If I'd been faithful, she would not have broken up with me!' (Both these statements are *counterfactual conditionals*. They are 'conditionals' because they are 'If...then...' statements. They are 'counterfactual' because the bit after the 'If', the antecedent, is false.) I tend to use statements like this when I am telling off my daughter: 'If you'd done your homework earlier, when you were less tired, it would have been easier!'

Enter Richard von Mises, mathematician (and scientist) extraordinaire, who devised one of the most detailed versions of the relative frequency interpretation of probability. One of his key ideas was that it is reasonable to *model* finite empirical collectives as infinite mathematical collectives. So probabilities concern infinite *mathematical* collectives (on which limit operations, discussed in the previous section, are well defined). But these mathematical collectives must have appropriate relationships to empirical collectives.

At first sight, this idea might seem a bit weird. But, as mentioned during the discussion of the subjective interpretation, in Chapter 5, idealizations are common in science, and especially in physics. And there is not normally any doubt that these idealizations represent real things. Infinity has a special role to play in many such idealizations. In ideal gases, molecules are *infinitely* small, point-sized, things. In optics, lenses are often considered to be *infinitely* thin. What's more, lenses' focal lengths are defined in terms of the images they would create of objects that were *infinitely* far away.

Von Mises uses an example from fluid mechanics to illustrate this point. (It requires an understanding of calculus, so I will not cover it. See Gillies (2000: 102–3) for an explanation, if you're curious.) Von Mises concludes:

> [T]he results of a theory based on the notion of the infinite collective can be applied to finite sequences of observations in a way which is not logically definable, but is nevertheless sufficiently exact in practice. The relation of theory to observation is in this case essentially the same as in all other physical sciences. (1928: 85)

This is a reasonable point. So let's accept it (for the time being, at least). Don't some of the other difficulties with the view that probabilities should be defined in terms of empirical collectives, which we discussed earlier, carry over? For example, how do we know that it is appropriate to model empirical collectives by using mathematical collectives *with limit values for relative frequencies*? After all, as we saw in the previous section, we can easily imagine infinite mathematical collectives where the relative frequencies do not have limit values. Think back to Hájek's (2009: 220) example of a sequence of coin flips:

HT HHTT HHHHTTTT HHHHHHHHTTTTTTTT...

However, von Mises anticipates this kind of objection. And it fails, according to him, *on two distinct empirical grounds*. Specifically, von Mises claims that there are two laws governing empirical collectives, which can be stated, roughly, as follows:

> (I) *The Law of Stability*: The relative frequencies of attributes in collectives become increasingly stable as observations increase.

(II) *The Law of Randomness*: Collectives involve random sequences, in the sense that they contain no predictable patterns of attributes.

We will consider these (alleged) laws in the next section.

4 The Empirical Laws: Stability and Randomness

Let's deal with the law of stability first. The idea is simple. Over time, the relative frequencies of attributes in empirical collectives fluctuate less and less, and tend towards specific values. Of course, this fits with the idea that there is a *limit* that these frequencies tend towards, at infinity, which we have already covered.

Von Mises holds that this law is confirmed by experience. So perhaps it is no coincidence that it was first proposed, in the sixteenth century, by a dedicated gambler who happened also to be a good mathematician. His name was Gerolamo Cardano, and he was responsible for the earliest work on probability (which was elementary in comparison to that performed in the seventeenth century). Presumably, he posited this law on the basis of his extensive experience as a gambler.

It is also interesting that this law was proposed *before* the laws of probability – the laws of addition and multiplication (explained in Appendix A) – were discovered by Fermat and Pascal. In fact, history suggests that belief in its truth was a prompt for their work. Recall Antoine Gombaud (a.k.a. Chevalier de Méré), the French gambler discussed in Chapter 2. He didn't *only* wonder about how to divide stakes when bets were not (able to be) concluded satisfactorily. He was also curious about his failure to win in a game that he had devised. A little background is necessary to understand why.

Gombaud made a good profit by betting repeatedly, at even odds (i.e. with a betting quotient of a half), that at least one six would appear in four rolls of a die. This is now unsurprising, because we can work out the probability of a six appearing, assuming a fair die. The probability of not getting a six, on each roll, is 5/6. So the probability of not getting a six four times in a row is $(5/6)^4$. Therefore, the probability of getting at least one six is $1 - (5/6)^4$, i.e. 671/1295. A betting quotient of a half on this event is unfair.

But later, Gombaud repeatedly offered a more complicated bet. (Perhaps he did so because his victims were becoming wise to the fact that the aforementioned game was in his favour.) This was that two sixes would appear together at least once, in twenty-four rolls of *two* dice. Again, he offered the bet at even odds.

Gombaud had a mistaken theory for thinking that this new bet would work just as well as his previous one. But he noticed that it didn't, and drew the matter to the attention of Pascal. He noticed, that is to say, that he began to lose over time. What makes this remarkable, however, is that the probability of his losing each bet was only lower than a half by a tiny amount, assuming fair dice. Again, we can now calculate it. The probability of not getting two sixes on one throw of two dice is 35/36. So the probability of not getting two sixes on twenty-four throws is $(35/36)^{24}$. And the probability of getting two sixes at least once is therefore $1 - (35/36)^{24}$, or approximately 0.4914. So Gombaud spotted a difference of just 0.086, between a half and 0.4914, through his repeated 'experiments'. What's more, to repeat, he spotted this despite *not* being a good mathematician. He had no way to calculate the value properly, which is why he consulted the mathematicians.

Now Gombaud would not have thought his findings to be interesting unless he *assumed* (implicitly or explicitly) that the relative frequency of wins would stabilize, and tend towards a specific value, pretty quickly. (It is hard to imagine that his total number of bets on this game was higher than ten thousand. It was probably of the order of one thousand.) And this assumption appears to have been made also by Fermat and Pascal. Think of it this way. If they had considered his results to be a freak occurrence, they would have had nothing to systematically investigate. So they must have been investigating (what they took to be) a stabilization value, and trying to understand why it was less than one half.

But is the law of stability *really* true? The way to argue from experience is typically just to give lots of examples of experimental results. You can find several by Googling 'relative frequency experiments'. Or you can do your own. Here is one that I conducted. I rolled a ten-sided die – a regular decahedron – four hundred times, and recorded the total number of tens rolled after each ten rolls. (Regular dice like this are used in many board games, including Dungeons and Dragons. We saw some other kinds back in Chapter 5.) I list some of the interesting data points, where some of the highs and lows of relative frequency occurred, in Table 7.4. (Relative frequencies are produced to three decimal places.) This is followed, in figure 7.1, by a graph showing how the relative frequency related to the number of rolls.

It was sheer luck that the experiment ended with a relative frequency of exactly 0.1, which is what we'd expect at

TABLE 7.4 Key data points in rolls of a regular decahedral die

No. of Rolls	No. of 10s	Rel. Frequency of 10s
30	0	0
60	5	0.083
70	5	0.071
90	9	0.1
140	11	0.079
180	20	0.111
220	27	0.123
400	40	0.1

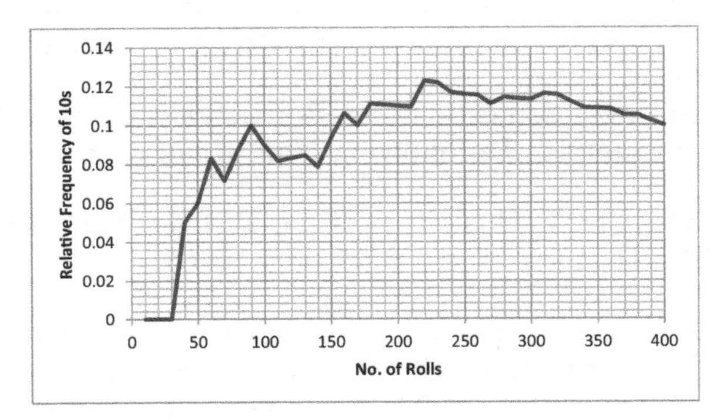

FIGURE 7.1 Relative frequency of 10s in rolls of a regular decahedral die

infinity. The relative frequency would have diverged again, if I'd continued rolling. (We know this for sure. Next, we will either end up with 401 rolls and 40 tens, or 401 rolls and 41 tens.) But we would not expect the relative frequency to ever again go as low as 0.071 or as high as 0.123.

Other than giving more examples of a similar kind, which would be extremely boring, there is not much that can be added, in order to empirically defend the law. (We have already mentioned the successes of casinos and insurance companies, etc.) At the very least, there seem to be many collectives that *do* obey the law (as far as we can tell, with the finite numbers that we can deal with). We could use some mathematical proofs in support – the so-called (strong and weak) *laws of large numbers* – but these presuppose that the sequences of variables in the collectives are *random*. So let's now move on to the law of randomness.

It is best to introduce randomness by appealing to intuitions about what's *not* random. And here's a clear example. Consider the following sequence of coin flips:

HTHTHTHTHTHTHTHTHTHTHTHTHT...

This is not random because there's a pattern in the attributes. And it's *obvious* that it's not random because it's *obvious* that there's such a pattern. Any gambler worth her salt would bet that a heads result would occur next. And then she'd bet on a tails result thereafter. And so on. She would exploit the lack of randomness of the sequence – the existence of a pattern of results – in order to formulate a successful betting strategy.

Patterns may also exist when it's much less obvious. Consider, for example, the following sequence of coin flips:

T HH T H T H TTT H T H TTT H T H TTT H TTTTT H...

Is there a pattern? You may be forgiven for thinking that there isn't. But there is. Heads results appear only at places 2, 3, 5, 7, 11, 13, 17, 19, 23, 29, and so forth. And you may now recognize these as prime numbers. So if you were to follow the rule of 'Bet on a heads result if the flip number is prime, and bet on a tails result if the flip number is not prime', then you would never lose a bet on the result of a flip.

What is the evidence that collectives are random? Consider how many gamblers try to devise systems for winning, without cheating. To this end, many gamblers record the results of previous games, e.g. spins of the roulette wheel, in the search for some kind of pattern. Indeed, I have seen some gamblers, in casinos, who spend the better part of their days doing this. (Naturally, recording the results can be used to determine the relative frequency of attributes too. So this might be used to spot if a roulette wheel is biased, due to being unlevel, for example. This *has* been done successfully. But that's *not* what we're discussing here. We are discussing the search for patterns in results by place number.) Are these gamblers successful? Do they ever succeed in spotting patterns that they can successfully exploit in order to systematically beat the casinos? It seems not!

This is not evidence that *no* patterns exist. It is simply evidence that they are extremely difficult to discern if they exist (on the basis of the data that gamblers are presently able to collect).

Recall now, for the last time, Hájek's (2009: 220) hypothetical infinite collective of coin flips:

HT HHTT HHHHTTTT HHHHHHHHTTTTTTTT...

We now see how von Mises would respond to any objection based on this. This collective violates not only the law of stability, because it has no frequency in the limit, but also the law of randomness, because there is a distinct pattern of Hs and Ts (which just happens to be simple to spot). But is this response good? Let's start to critique von Mises's version of the relative frequency view by considering this, in dialogue format.

5 Initial Criticisms of Hypothetical Frequentism

DARRELL: OK, who is going to be first to argue against von Mises?

STUDENT ONE: You may have a fight on your hands!

DARRELL: That would be better than silence. Why don't you go first?

STUDENT ONE: OK. The big plus point of considering *actual* finite frequencies is that they're easy to measure. That is, although we saw the problems with thinking that probabilities are identical to relative frequencies in finite empirical collectives...

DARRELL: So your idea is that in trying to solve the problems, the plus point has been lost?

STUDENT ONE: Yes! Lost, or at least *diminished*. Let me put it this way. Consider all coin flips. And imagine we have lots of data on this collective. For example, let's say we have the results of all coin flips made this year. Should we be confident that this data is relevant to a probability concerning the results of coin flips?

STUDENT TWO: Yes, because of the empirical laws?

STUDENT ONE: But that's just it. I am not convinced that they *are* general laws. Or, at least, I think it's *reasonable* not to believe that they are. First, perhaps there is *no* frequency in the limit for the results of coin flips. Second, even if there *is* a frequency in the limit, perhaps it is significantly different from what our present data leads us to expect. Third, perhaps there is a pattern in the results of coin flips that we cannot spot, which renders it non-random. Maybe every ten millionth flip is heads, or some such. We just haven't got enough data to spot that pattern!

STUDENT TWO: That all seems a bit sceptical to me. I mean someone could try to shoot me today, but I'm not going to wear a bulletproof vest...

STUDENT ONE: Then let me argue in a slightly different way. Let's consider the third issue, about randomness. Let's say you're right, that we have evidence that the sequence of coin flip results *is* random. Now imagine that, in the distant future, a pattern in the sequence is discovered. Would you *really* want to accept that we had been wrong – even if we had been perfectly rational – to use probabilities concerning coin flips for all the time before?

STUDENT TWO: That's a better point, I grant you. And I also have to confess to doubts that the laws are *fully* general, even if they are *typically* true of empirical collectives. Indeed, maybe the laws are *probabilistic* in character!

STUDENT THREE: There is something along these lines that bothers me. Wouldn't it be weird if these laws *just so happen to be true*? *If* they are, doesn't this need some kind of explanation?

DARRELL: That's a good question. In fact, the next interpretation of probability we'll consider – the propensity interpretation – tries to explain why we should expect stability and randomness in some cases. Let's not go into the details now. Let me instead suggest the following. Von Mises would have replied to you that these laws are just brute facts, like the law of conservation of energy in physics. And he would have added that if there are no unexplainable features of the world, then we require an infinite regress of explanations.

STUDENT THREE: Well I have a reply to that, as it happens...

DARRELL: Yes?

STUDENT THREE: It relies on the distinction, drawn by Carnap, between *empirical* and *theoretical* laws. Empirical laws operate at the level of observable things, whereas theoretical laws operate at the unobservable level. Now often, in physics, the empirical laws can be *explained* by the appropriate theoretical ones.

DARRELL: Examples being?

STUDENT THREE: Consider 'All iron bars expand when heated'. There's an explanation for that in terms of the constituents of iron bars, and what temperature really amounts to, namely average velocity of molecules.

DARRELL: Good. So your idea is that these laws about collectives are *laws about observables* for which we should expect underlying *laws about unobservables*, in the same way we normally do in physics?

STUDENT THREE: Exactly. So von Mises's analogy with physics is not as good as it first seems. Instead, something is missing from his account of probability. Roughly, I'd say it focuses on the superficial, the observable.

DARRELL: But some philosophers are a bit suspicious of talk about unobservable things – so they might see this 'superficiality' as advantageous.

STUDENT THREE: Maybe. But most of those people still think it's OK to have theories with unobservable things in them. They just don't *always* take the theories to be true, or highly truth-like.

DARRELL: Point taken.

6 More Criticisms of Hypothetical Frequentism: Single Cases, Reference Classes, and Sequence Ordering

DARRELL: OK. Let's move on from focusing on the alleged laws. Can anyone think of any other problems with so-called 'hypothetical frequentism'?

STUDENT TWO: My main worry, which you touched on when you introduced frequency interpretations, is that it doesn't allow for probabilities concerning single events – or probabilities concerning individual things in 'mass phenomena' collectives.

DARRELL: And why do you think that these are important?

STUDENT TWO: I'll give an example. Imagine I am in hospital. I am seriously ill, and the consultant recommends an operation. I ask her for a prognosis. I ask for the probability that I'll survive the recommended operation.

DARRELL: And her reply is?

STUDENT TWO: '70 per cent.' But because I know a bit about the interpretation of probability, I push her on what she means by this. Is it just her opinion?

DARRELL: And her reply?

STUDENT TWO: 'It's based on medical research.' And so I ask 'What's the relevant collective?' or, as I've also heard the question posed, 'What's the reference class?'

DARRELL: OK, so she then tells you about the research. Maybe it involved two hundred patients at a particular hospital, operated on by several different surgeons, who used different surgical techniques, and so forth.

STUDENT TWO: Right! So I then declare 'But I want to know the probability that *I* will survive!' I continue 'It must be different from the one you've given me, because this is a different hospital, you are a different surgeon, and you propose to use a specific surgical technique...'

DARRELL: Ha! I have been in some similar situations. But if the consultant is a true frequentist, she must respond that there is no world-based probability.

STUDENT ONE: Is this really so bad? Even if there *were* a probability for just you, the patient in those circumstances, surely we wouldn't have decent data on it, anyway? To get that, we'd need to repeat the same scenario multiple times – and that seems impossible, when the operation is irreversible.

DARRELL: Good point. But maybe there are some other cases, similar in kind, where we do want to say there is a probability. How about a single radioactive atom, such as an atom of carbon-14? Does *it* have a probability of decaying over some arbitrary period of time?

STUDENT ONE: Why not just say 'No'? It's radioactive atoms *of that type* that have half lives, and hence decay probabilities, *when considered as collectives*.

DARRELL: I am not so sure. Think of the following thought experiment. There is just one radioactive atom of type x produced ever, in the whole universe. Does it...

STUDENT ONE: I get it...

DARRELL: Then I will say only this. It seems like your position commits you to the view that whether the single atom of type x decays or not, over some period of time, is physically *determined*. That is, if you insist that there is no world-based probability concerning it in isolation.

STUDENT ONE: Sure.

STUDENT TWO: But do we know that? I mean, shouldn't we be willing to keep our options open on whether the world is indeterministic?

DARRELL: I am inclined to agree.

So far, the dialogue has covered the problem of single case probabilities. But there is a related problem, known as the *reference class problem*, which it will now cover.

STUDENT THREE: OK. I have a different objection, if we can move on?

DARRELL: Sure.

STUDENT THREE: Go back to the example of Student Two in the hospital. Let's accept that there's no *individual* probability. There's still a problem about *which* collective to consider.

DARRELL: Can you explain further, perhaps by modifying the thought experiment?

STUDENT THREE: Sure. Imagine the consultant knows of *two* studies on the type of operation she intends to perform, which report *different* relative frequencies for patient survival. Which study should be used?

STUDENT TWO: This is known as the *reference class problem*, right?

DARRELL: Yup.

STUDENT TWO: We can make it more interesting by making the studies relevant to me, the patient, in different ways. Let's say Study One considers patients operated on in the same hospital that I am in, over an age range of 10–80. And let's say Study Two considers patients operated on in a different hospital, but over an age range of 20–30. I'm 23, by the way.

STUDENT THREE: Right. So maybe Study One had a 50 per cent survival rate, whereas Study Two had a 70 per cent survival rate. Which to use?

STUDENT ONE: Isn't there a simple solution? Shouldn't we combine the results, and look at the relevant frequency of survival over *all* the patients in the studies?

STUDENT THREE: Not so fast! There's an argument for using Study One in isolation, namely that the operations were performed in the same hospital. But there's also an argument for using Study Two in isolation, because it doesn't include patients much younger or older than me.

DARRELL: Right. And it seems reasonable to prefer to rely on one study, although it's a difficult task to say which should be preferred. Maybe the quality of the hospital in Study Two was inferior. Perhaps it was less clean, had less up to date equipment, and so on. And maybe the old age of some of the patients in Study One led to the low survival rate.

STUDENT ONE: I see. And actually, this is linked, I think, to the problem of single case probabilities.

DARRELL: That's an interesting idea. How so?

STUDENT ONE: Ideally, we would not want to use either study, because the collectives are too broad. We'd want one done in the same hospital. With people of the same age as Two. And the same sex as Two. And the same medical history as Two. And the same DNA as Two... so, really, we'd want a study on Two!

DARRELL: Indeed...

STUDENT TWO: And even if we accept that we can't have that, we don't always have a solid way to judge which of two collectives is better to use.

DARRELL: Right. One of the reasons is that *similarity* is being judged. To illustrate, consider a steel sphere and a wooden pyramid. Which is more similar to a wooden sphere? The answer is plausibly a matter of context. If we're thinking about composition, ductility, or thermal conductivity, we'll pair the wooden items. If we're thinking about shape, surface area to volume ratio, or disposition to roll, then we'll pair the spheres.

We have covered the two main objections to hypothetical frequentism, and seen that they are somewhat related. But there is one final objection to cover, which is presented by Hájek (2009). He calls it the *reference sequence problem.*

The basic idea is that the frequency in the limit of an attribute in an infinite collective depends on the *order* of the members of the collective. To see this, consider two infinite collectives with the same members, namely all the natural numbers. The first collective is ordered as we normally count: 1, 2, 3, 4, 5, etc. But the second is ordered rather differently: 1, 3, 2, 5, 7, 4, 9, 11, 6, etc. Now consider the relative frequency of even numbers. In the first collective, this is a half. In the second collective, this is a third.

One response is to insist that only one of the ordering strategies is 'natural'. But this is implausible, even when we're dealing with simple events like coin flips. The 'natural' order for coin flips is temporal; if flip *a* occurs before flip *b*, then the results of flip *a* should appear earlier in the sequence than those of flip *b*, and vice versa. According to special relativity, however, whether *a* occurs before *b* is a matter of perspective, i.e. the frame of reference of measurement. So in one frame of reference, *a* occurs before *b*. In another, *b* occurs before *a*. And in a third, *a* and *b* occur simultaneously. This leads to the weird situation where the relative frequency of 'heads', for the flips of some coin, can be different for two observers.

But how about appealing to randomness, instead? (The example sequences of natural numbers, above, are not random.) Then the problem is that the same set of events may be random for one observer, but not for another. (Imagine, for instance, that an observer randomly changes velocity, and often moves at extremely high speeds.) So what are we to say? The conclusion that there are probabilities involving those events for one observer, but not the other, is highly counterintuitive. We want to say, rather, that there are *observer-independent* world-based probabilities, if indeed there are any world-based probabilities present.

7 A Brief Sympathetic Conclusion

There is clearly something good about the motivation behind the frequency interpretation; to base probability theory on intersubjectively observable phenomena, namely frequencies. As we have seen, however, it is very difficult to defend the view that probabilities are simply *actual* frequencies. So a retreat is forced, where actual frequencies become fallible *indicators* of probabilities – e.g. frequencies in the limit, or hypothetical frequencies – rather than probabilities in their own right. In a nutshell, we might say that the focus on actual frequencies as the only empirical evidence *for* world-based probabilities is correct. But we might seriously doubt that frequencies *are* probabilities, even when these are construed as hypothetical. We might also question the value of insisting that world-based probabilities must be construed as non-actual frequencies, if they are not to be construed as actual frequencies. After all, non-actual frequencies are *not* directly measurable. And this brings us to the next chapter.

Further Reading

Although the frequency view remains popular outside philosophy – e.g. among statisticians – it is not the subject of much, if any, active research. We have already considered some arguments from the key critical papers in recent years, namely Hájek (1997) and Hájek (2009), which are intermediate to advanced in difficulty. Slightly more accessible treatments appear in Kyburg (1970: ch. 4) and Gillies (2000: ch. 5). Von Mises (1928) is reasonably accessible and worth revisiting.

8
The Propensity Interpretation

In the last chapter, we saw that *the law of stability* holds for many empirical collectives. That's to say, the relative frequencies of attributes in many collectives fluctuate less and less as observations increase in number. We saw this in the experiment where I repeatedly rolled a regular ten-sided die, for instance. The relative frequency of 'ten' results fluctuated in value increasingly less, and approached the value of one-tenth.

But *why* does the law of stability hold? Or more accurately, why does it hold for *some* collectives, but not for others? What makes *those* collectives special? The frequency interpretation of probability provides no answer to these questions. And many of its advocates would strongly object to the idea that such questions require an answer. Von Mises, for example, thinks as follows. The theory of probability is an empirical science, concerning mass *phenomena* (i.e. observable things), and has no place for speculation about underlying mechanisms (or 'metaphysics').

Advocates of the propensity view, however, seek to explain the law of stability – and in some cases, the law of randomness – in terms of more fundamental, underlying, things. And they take the underlying things – the propensities – to be the *real* probabilities. So on this view, probabilities generate, *but should not be identified with*, stable relative frequencies.

An analogy might help at this juncture. Think of the ideal gas law, also known as Boyle's law, which holds approximately for many real gases, such as helium. It relates observable properties, like volume, pressure, and temperature. But there is a deeper microscopic explanation of those relations, in terms of the molecules of the gas and *their* properties (such as mass and velocity). The so-called 'ideal gas law' is a result of *mechanical* laws governing the (ideal) gas molecules. Temperature is linked to the kinetic energy of the molecules, and so on. That's what contemporary physics says.

Similarly, in the current context, *frequencies* are the observable things. And *propensities* are the unobservable things. However, there is a significant way in which the analogy is imperfect. Many successful predictions have been made from the theory that gases are composed of molecules. The theory of propensities, however, only seems to *explain* the features of the sequences of attributes in some empirical collectives. We will come back to this issue.

1 Probabilities as Dispositions

All versions of the propensity theory – there are many, of which we will consider several, and this results in unavoidable confusions – have one thing in common. They rest on the idea that probabilities are *dispositions*. Consider words like 'soluble', 'flammable', 'fragile', 'sonorous', and 'ductile'. These correspond to ways that things are disposed to act under certain circumstances, or *dispositional properties* that things can possess. A bell may be sonorous in so far as it *would* sound when struck, even if it is never, in fact, struck. And similarly, a wine glass may be fragile in so far as it *would* shatter if it were dropped onto a stone surface from a height greater than 30 cm, even if it is never, in fact, dropped in such a way. Closely related to the idea of a disposition is that of a *power*. For example, we may say that pure water *has the power* to dissolve sodium chloride just as we may say that sodium chloride *has the disposition* to dissolve in pure water.

A little reflection makes it clear how important dispositional properties are in our descriptions of things. You are probably in a room (or other inside space) with windows, as you read. And you probably think that the windows are made of glass. But now imagine that a strongman struck one of those windows as hard as they could, with a sledgehammer, and failed to break it. What would you think? You would suspect that the window was *not* made of glass, in so far as it had been shown to *lack* the disposition that you thought it did. That is, assuming you were satisfied that the sledgehammer was genuine, and so on.

You need not assume anything about the sledgehammer (and so forth). More generally: If you are convinced that *A* and *B* have the disposition to make *C* occur, then you will infer that at least one of *A* and *B* is absent when *C* does not occur. So if you see a white crystalline solid added to a colourless liquid, and none of the white crystalline solid dissolves, then you might conclude that the statement, 'The white crystalline solid is sodium chloride and the colourless liquid is pure water', is false. This is a much more natural initial reaction than doubting that sodium chloride *really* has the disposition to dissolve in pure water.

2 Single Case Propensities (Popper)

One of the key problems with the relative frequency view is that it leaves no place for world-based probabilities concerning single cases, such as one-off events. There is no probability that the Allies would have defeated Germany by Christmas 1944, in World War II, if Operation Market Garden had been successful. (Watch *A Bridge Too Far*, if this means nothing to you. It's a great film!) There is no probability that Anne Boleyn would not have been beheaded, at Henry VIII's behest, if she had given birth to a baby boy. And so on. There are no such probabilities in history at all! There are no such probabilities even in simple cases like individual rolls of dice, or flips of coins, even though these are more easily thought of as parts of large collectives. So, as we noted in the last chapter, anyone who insists on admitting that there are probabilities concerning single cases must be a pluralist, and take them to be information-based, if they are committed to a relative frequency world-based view of probability.

Popper's version of the propensity theory was intended to avoid this problem. My own way of thinking of this is as follows. First, it is possible to imagine dispositions that are *disjunctive* (or 'either...or...') in character. Imagine when substance X comes into contact with substance Y, one of two things always happens: either X turns into lead, or X turns into pure gold. Then we may say that 'X is disposed to transmute into lead *or* to transmute into pure gold, when in comes into contact with Y'. We may also imagine that each happens some of the time. There is nothing obviously contradictory about this. And there's nothing obviously wrong with thinking that no other factors affect what happens when X is added to Y. That's to say, it is possible to think there is no further factor (or group of factors), Z, that is present *only* in the cases when X turns into lead (or gold).

Second, it is possible to imagine that the *disjuncts* – the two items connected by 'or' – have different weights. For example, there could be *more of a tendency* for X to transmute into lead than there is for X to transmute into gold, when added to Y. My own preferred way to think of this – I am not sure if it is anyone else's – is as follows. *Conflicting dispositions* can be present, and one disposition can be stronger than the other. The disposition to turn into lead may be stronger than the disposition to turn into gold, or vice versa, to a specific degree. On this view, *both* of the following can be true: 'X is disposed to transmute into lead when it meets Y' and 'X is disposed to transmute into gold when it meets Y'. We can add 'X is *n* times more disposed to transmute into lead than it is to transmute into gold, when it meets Y'.

You may have the following worry. When we initially considered dispositions, above, one of the examples was 'Sodium chloride is disposed to dissolve in pure water'. And when we say that, don't we mean that it's disposed *only* to do that? The response I have in mind is that what we *really* mean, and ought to say more precisely, is that 'Sodium chloride is disposed *only* to dissolve when added to pure water'. I omitted 'only' several times previously. In the discussion of disjunctive dispositions, for example, I should have written: 'X is disposed *only* to transmute into lead or to transmute into pure gold, when in comes into contact with Y'.

On such a view, probabilities are possessed by particular physical states of affairs, such as experimental setups, rather than collectives: 'admissible sequences must be either virtual or actual sequences which are *characterized by a set of generating conditions* – by a set of conditions whose repeated realization produces the elements of the sequences' (Popper 1959a: 34). Now, we can pull the key trick, claims Popper, which allows us to posit single case probabilities: '[S]ince the probabilities turn out to depend upon the experimental arrangement, they may be looked upon as *properties of this arrangement*' (1957: 67). Thus, according to Popper, '[A] singular event may have a probability even though it may occur only once; for its probability is a property of its generating conditions' (1959a: 34).

But does this *really* follow? At some points, von Mises, whom we encountered in the previous chapter, appears to endorse the view that world-based probabilities should be associated with physical setups (or states of affairs), *despite* his denial of the existence of single case probabilities. For example, he writes:

> [F]or a given *pair of dice* (including of course the total setup) the probability of a 'double 6' is a characteristic property, a physical constant belonging to the experiment as a whole... The theory of probability is only concerned with relations existing between physical quantities of this kind. (1928: 14)

So von Mises's view is not as dissimilar from Popper's as it first appears. A reason for being cautious about the existence of single case propensities will become apparent in the next section.

3 Single Case Propensities vs. Long Run Propensities

As we saw above, in section two, von Mises *agrees* with Popper that probabilities can be physical properties, but *denies* that such probabilities can exist in single cases, i.e. relate to outcomes in individual experiments, rather than to (collectives of) outcomes in repeats of the same kind of experiment. But why might that be?

One reasonable answer is that stable long run frequencies do *not* necessarily indicate the presence of single case propensities, and can be *explained* without appeal to these too. So presupposing that single case propensities exist goes considerably beyond the phenomena – the appearances – and into the murky realm of metaphysics. Consider the following, in support.

Imagine that an experiment with two possible outcomes is performed infinitely many times, and produces relative frequencies, f and $1 - f$, for each possible outcome. (The 'infinitely many times' is used to rule out the possibility of the relative frequencies diverging from the relative frequencies in the limit. If you think such a scenario is impossible, instead imagine that the experiment is performed an *extremely* high number of times, and that the law of stability, discussed in the previous chapter, holds. Hence, f and $1 - f$ are approximately equal to frequencies in the limit.) Imagine also that the experiment is performed well. Should we conclude that f and $1 - f$ represent the single case propensities for each possible outcome *in any individual run of the experiment*? This is a (simplified version of a) question posed by Paul Humphreys (1989: 52).

It seems not, Humphreys correctly concludes, unless we have further information. And that's because the scenario as described is compatible with (at least) two different underlying situations, corresponding to two different senses in which the experiment can be 'performed well' (i.e. in accordance with its design).

First, imagine that the experimental system is *indeterministic*, and that all the factors causally influencing the experimental outcome are held constant across iterations. In other words, all of the initial conditions relevant to the applying the complete laws of nature governing the experiment's outcome are *identical* in each repeat. Hence, the probabilities *appear* in the complete laws of nature (relevant to describing the experimental outcome). And since the laws are *complete*, the use of probabilities cannot be avoided by appeal to further laws. Consider the following specific example, to illustrate. A die is rolled in exactly the same way, each time. It begins in the same orientation (i.e. position) relative to the rolling mechanism. The rolling mechanism works in exactly the same way each time (i.e. it applies the same forces to the die). The die lands on exactly the same surface. And there is no outside interference. That's to say, the experiment involves a 'closed system', such that no external forces or factors influence the outcomes. Now if the outcomes were not always the same, we would correctly take the system to be *indeterministic* (provided we were *convinced* that the system was closed, and that the outcome was determined only by the forces setting the die in motion, its initial position, and the properties of the surface it landed on). It may help to take another look at the discussion of Laplace's demon, in Chapter 1, if this is not entirely clear.

Second, imagine that the experimental system is *deterministic*, but that some of the factors causally influencing the outcome are randomly varied across iterations. Then the outcomes of each experiment may differ, despite the outcome in each case being uniquely determined by the laws of nature and the initial conditions. For illustrative purposes, the die rolling example given in the paragraph above may be altered in just one respect; the initial orientation of the die may be allowed to vary randomly (although all other aspects – the rolling process, and so on – remain fixed). Now if there were different outcomes – it would be extremely surprising if there weren't – then it would be wrong to conclude that single case propensities were present. *Nonetheless, the experiment is designed in such a way that it will result in characteristic frequencies of outcomes, when it is repeated a large number of times.*

This second example illustrates the possibility of what Jim Fetzer (1988) and Gillies (2000) call *long run propensities*. We'll adopt the latter's definition (which is slightly different from the former's, because it doesn't involve limits or infinity):

> A long-run propensity theory is one in which propensities are associated with repeatable conditions, and are regarded as propensities to produce, in a long series of repetitions of these conditions, frequencies which are approximately equal to the probabilities. (Gillies 2000: 126)

Indeed, it seems as if von Mises had something like these in mind, given his belief in the law of stability, when making his aforementioned comments about 'physical quantities'. Let's now explore how exactly these relate to *single case propensities*, if at all, in a dialogue.

4 Single Case and Long Run Propensities: What's the Relationship?

STUDENT ONE: I've been reading some of Popper's original stuff on propensities, and it seems to me like it's a bit confused. To be specific, it seems like Popper's view isn't entirely single case or entirely long run. It's some sort of weird admixture of the two views you've presented.

DARRELL: I think you're right. Can you illustrate this with some quotations?

STUDENT ONE: Sure. He writes: 'propensities turn out to be *propensities to realize singular events*' (1959a: 28), but also that a repeatable situation, like an experiment, can have a 'propensity to produce sequences whose frequencies are equal to the probabilities' (1959a: 35). Also, in his earlier piece, he wrote: '[The] *properties of the arrangement* . . . *characterize the disposition*, or *the propensity*, of the experimental arrangement to give rise to certain characteristic frequencies *when the experiment is often repeated*' (1957: 67).

DARRELL: Nicely done. Now can anyone think of a charitable way to read this? Could there be a way to resolve the tension between these apparently conflicting statements?

STUDENT TWO: Hmm. I wonder if the presence of single case propensities might *entail* the existence of long run propensities, in repeats of relevantly similar cases.

STUDENT ONE: That's an interesting thought.

DARRELL: Can you explain further, maybe with an example?

STUDENT TWO: Your wish is my command. Imagine we have a situation where indeterminism holds; that's to say, where the outcome of the situation is not uniquely determined by the initial conditions. Let's say it is a quantum mechanical system, where there's an equal single case propensity – of one half, of course – for each of two possibilities, *A* and *B*, to occur.

STUDENT ONE: Duly imagined. I'm thinking of spin measurements, for those who know the physics.

STUDENT TWO: OK. Now let's imagine that *exactly* the same kind of situation were to be set up repeatedly. In the long run, what will be the relative frequency of *A*?

STUDENT ONE: The same as – or, I should say, approximately the same as – the underlying propensity for *A* in that situation.

STUDENT TWO: Right! So those repeatable conditions have a long run propensity *because of* the single case propensity present in every single case in which those conditions obtain. Boom. We have long run propensities on the back of single case propensities.

STUDENT ONE: Congratulations! You have saved Popper from himself.

DARRELL: Let me be the first to raise a doubt. Might there not be *some* situations where a single case propensity is present but the conditions are *not* repeatable?

STUDENT TWO: How do you mean?

DARRELL: I am thinking of David Miller's (1994) account of single case propensities, according to which they depend on the *whole state of the universe* (or something close). If that's right, then there's a sense in which the conditions cannot be repeated.

STUDENT TWO: Well they *may repeat* in different universes, right? And there'd be long run propensities across those conditions in different universes (where the same laws of nature hold)?

DARRELL: That's true, but this is highly speculative, and detached from experience. So maybe I can address the doubt I mentioned – the doubt that Miller would raise – somewhat differently.

STUDENT TWO: How?

DARRELL: Single case propensities for particular conditions need only be associated with what's causally relevant to the outcome, in the universe. And that can be a lot less than the whole universe. Here, I follow the view of single case propensities put forward by Fetzer, according to which: '[P]ropensities for outcomes...depend, in general...upon a complete set of (nomically and/or causally) relevant conditions...' (1982: 195).

STUDENT TWO: Cool. So are there any other problems with what I suggested?

STUDENT THREE: Yes. You can't *define* world-based probabilities as *both things – single case and long run – at the same time*.

STUDENT TWO: Agreed. So my suggestion is as follows: Popper was trying to define them as single case – or at the very least, should have defined them as single case – and to show that long run propensities, and hence appropriate relative frequencies, would result. In summary: a world-based probability is a single case propensity, which results in a long run propensity, and hence appropriate relative frequencies in the long run.

DARRELL: That *is* a charitable reading, and a useful one even if it's too charitable.

STUDENT THREE: OK. But then, as we have already seen, Popper appears to miss a trick. That's to say, he misses – at least, in the early works on propensities we're referring to – the possibility that some world-based probabilities might be long run propensities *without being single case propensities*.

STUDENT ONE: Yes, that seems right...

STUDENT TWO: It is. I agree. Darrell's deterministic die rolling example was like that; the long run propensity was in the set of repeatable conditions, but there were no single case propensities present in any particular instance.

STUDENT THREE: So why not define probabilities as long run propensities, as that will catch *all* cases where stable frequencies arise?

DARRELL: Well, one does seem to lose world-based probabilities in single cases.

STUDENT TWO: Right. So here's what I think. I think world-based probabilities can be *either* single case or long run.

STUDENT THREE: But how are we supposed to tell which is which, in any given case? I mean we can't just look at the stable frequencies, the observables, and conclude that single case propensities exist.

STUDENT TWO: That's true. But we can make sense of their existence, even if it's ultimately right that we can't empirically confirm that they exist. I am just saying that we should keep an open mind.

STUDENT THREE: I am thinking of science. I am trying to keep our definition grounded in experience, and to keep it simple – that is, so we don't need different world-based definitions in different cases. We can keep in mind the possibility that the universe is, or particular systems therein are, indeterministic. But we can do that while defining world-based probabilities as long run propensities *in general*. Indeterminism is then just *one mechanism by which* such probabilities can come about.

DARRELL: This is a nice debate – Student Three, I think you're arguing for a view rather like Gillies's (2000) – but I'm going to have to stop it there.

5 The Reference Class Problem Again

STUDENT ONE: Wait! Before you continue, I have a burning question. Doesn't the reference class problem, or something similar, arise for propensities, just as it does for relative frequencies?

DARRELL: Perhaps. Can you be more specific?

STUDENT ONE: Sure. Think of it this way. Single case propensities are said by many to be *conditional* on particular physical setups, just as relative frequencies are conditional on specific collectives. For example, Popper (1967: 38) writes of '*properties of the repeatable experimental arrangement*'. But when, exactly, is an arrangement, or indeed an experiment, *repeated*? Doesn't that depend on how we describe it?

STUDENT TWO: I think I get it...But I'd like to check. Mind if I give an example to see if I do?

STUDENT ONE: Fire away.

STUDENT TWO: Imagine a scientist repeatedly measures the potential difference across a particular component in a specific circuit. He uses one kind of potentiometer to do the measurement ten times in a row. He then uses a different kind of potentiometer to do the measurement ten times in a row. Are all the experiments the same? Or are the first ten experiments different from the next ten?

STUDENT ONE: Nice example. You hit the nail on the head. But the problem is more pressing than your thought experiment illustrates. Imagine now that the equipment and technique used in measuring the potential difference are exactly the same for each of the twenty measurements. But think about the temperatures of the lab, and indeed the circuit. It's unavoidable for these to fluctuate very slightly, even if the experimenter is extremely careful – even if a good thermometer is used to check. There are always limits to precision! So are the experiments *the same*?

STUDENT THREE: You're taking things too far. After all, it is perfectly normal, as part of an experiment, not to control for *all* the factors that can vary. Indeed, the point of many experiments is to let *some* factors vary, while holding others fixed – or, if you insist, *approximately fixed* – in order to test whether the varying factors have any causal influence on the results.

DARRELL: That's true, but I still think there's a real issue here.

STUDENT THREE: Maybe. But it doesn't seem to me to be a problem for the propensity account of probability, any more than for scientific method in general. 'When are two experiments of the same type?' is a question that can be asked – and is asked – even when probabilities aren't being used.

DARRELL: Fair point. And maybe we should use the definition of single case propensities offered by Fetzer, which I mentioned earlier, to help with thinking when experiments are suitably similar to indicate the existence of such propensities. Specifically, we could say that experiments are 'the same' *in the sense that they may be used to measure single case propensities* only when the conditions causally relevant to the outcome(s) are invariant across those experiments.

STUDENT THREE: We could say that. But maybe we could relax that requirement somewhat.

DARRELL: You have in mind that a little variability in the conditions can be OK? That is to say, it's enough for the single cases propensities to be approximately equal, in each case?

STUDENT THREE: Yes. Again, many real experiments are like this, even when they don't involve propensities. We know that differences in temperature can affect the resistance of circuit components, as mentioned above. But we don't worry about small differences, when we measure such resistances, because we know the effect that they have is negligible.

DARRELL: OK. But we are still left with the deep methodological problem – really, this is the one you raised in the last discussion – even if it isn't the *reference class* problem. Think of it this way. Imagine we fix all the factors we know to be causally relevant, across experiments, and notice that outcomes vary nonetheless. Do we conclude that there's a single case propensity, which we're measuring? Or do we conclude that there's some causal factor that we haven't taken into account, which is varying across those experiments?

STUDENT THREE: Yes. Science is a messy business, especially when complex systems are involved. That's why I think it's safer to use long run propensities, which can cover both cases, by default.

6 A Final Objection to Probabilities as Single Case Propensities: Humphrey's Paradox

Before we conclude, let's cover a final objection to the single case propensity account of probability. Think back to the discussion of dispositions in single cases, in the first section. When we say a bullet striking a window (at high speed) has the disposition (only) to shatter it, do we not mean that a bullet will *cause* a window to shatter when it strikes it? And might we not represent that as a single case propensity with a value of unity? It seems so; it seems we may say 'P(*window breaks*, *bullet strikes the window*) = 1'. But then single case propensities of lower positive values are like watered-down causes. As Popper suggested: 'Causation is just a special case of propensity: the case of a propensity equal to 1' (1990: 20).

There's a problem with this, however, which was first noticed by Humphreys. And once you see it, it is obvious. The trick is as follows. If P(p, q) is well defined, then P(q, p) is well defined. But this appears to mean that if we treat P(p, *q*) as a single case propensity, then we should treat P(*q*, *p*) as a single case propensity. And this is odd because causation operates in only one temporal direction. To see the point, consider P(*bullet struck the window*, *window breaks*). Imagine we see a window break, and work out that the value for the aforementioned probability is 0.5. Surely we couldn't understand this as a single case propensity. It is not as if windows have some kind of bullet attraction property.

The long run propensity theory can avoid this problem by appealing to the repeatable conditions, and what would happen in the long run. For example, imagine we are considering windows *in cars driven in a specific war zone*; there's a reasonable long run propensity for those windows to break by being hit by bullets. If we consider windows *in cars driven in countries where guns are illegal and extremely rare*, by way of contrast, the long run propensity for breaking by gunfire is much lower. So we should really write of probabilities such as P(*bullet struck the window*, *window breaks while in a car driven regularly in Baghdad in 2010*), versus P(*bullet struck the window, window breaks while in a car driven regularly in the UK in 2010*), and so forth. (The UK has strict gun laws, by the way!)

The situation for the single case propensity view is, however, rather more dim. As a result, Humphreys (1985) himself takes his paradox to show that single case propensities are *not* probabilities, although they nonetheless exist. Fetzer (1981), on the other hand, defends the view that although single case propensities do not satisfy the standard axioms of probability, they are probabilities nonetheless. He promotes a broader view of probabilities, as not being limited to any one set of axioms.

7 A Brief Conclusion on Propensities

Overall, we have seen that the long run propensity view of world-based probabilities suffers from fewer criticisms than the available alternatives (i.e. the relative frequency or single case propensity views). However, we should not be too quick to conclude that there are no single case propensities, or to deny that probabilities might be understood as single case propensities in some special cases. Moreover, there are other subtle alternatives explored in some of the literature mentioned below.

Further Reading

Many different versions of the propensity interpretation are discussed in the contemporary research literature; the following are all at an advanced, or intermediate to advanced, level. On the constructive side, for example, see Gillies's (2000: ch. 7) work on developing the long run version in a non-operationalist way, and showing how it may be used to derive the empirical laws of probability. On the destructive side, see Eagle's (2004) extensive enumeration of arguments against propensity theories, only some of which we have discussed here. Handfield (2012) is highly recommended for getting to grips with the state of the art on propensities. Finally, Suárez (2013) has more recently argued for a pragmatic view on which propensities are responsible for, but should not be *identified* with, world-based probabilities.

9
Fallacies, Puzzles, and a Paradox

We have now covered all of the interpretations of probability, and many variants thereof. So it's time to examine the fruits of our labour. In this chapter, we will see how the interpretations may be used to shed light on some common fallacies, related puzzles, and an entertaining paradox. Learning about these is of independent value, in improving your reasoning with probability (and avoiding some surprisingly common mistakes).

1 The Gambler's Fallacy and the 'Law' of Averages

Let's begin with a well-known story. In 1913, at Le Grande Casino in Monte Carlo, something extraordinary happened at one of the roulette tables. (If you've forgotten about roulette, look back at section six of Chapter 4.) The ball landed in a black compartment time and time again. In fact, it landed on 'black' twenty-six times in a row.

How do you think the gamblers bet at that point? How, indeed, do you think they'd been betting, in increasing numbers, for the last six spins of the wheel? (The news had spread that there had been so many black results in a row. Gamblers were crowding around the table as a result.) The answer, which you will hopefully find surprising, is that they were betting on *red*!

What were they thinking? The short answer seems to be this. They continued to assume that the wheel was fair. And they concluded that a red result must be due next, with increasing probability. After all, the temporary imbalance would have to be offset later, yes? Isn't that what 'the law of averages' says? What's more, the probability of twenty black results in a row is incredibly low. The probability of twenty-one black results in a row is lower still! So it is really improbable that anyone would ever witness such an event.

This is a seductive, but utterly wrong, way of thinking. Why? The gamblers assumed that what would happen on the next spin of the wheel *depended* on what had happened before. In truth, however, the result of each spin was *independent* of any other spins. It *is* extremely unlikely that a fair roulette wheel would give twenty black results in a row (let alone twenty-six). So you would be right if you expected never to see such a thing happen. But it does not follow that there is a low probability of a fair roulette wheel giving twenty-one black results in a row *when it has already given twenty black results*. It's fair! So the probability that it gives a black result, on any given spin, is one half. Or to express matters in more formal terms: P(black) = P(black, the last n results were black) for any value of n.

If we think in terms of world-based probabilities and the law of stability – which we discussed in the last two chapters – we can see another mistaken way in which the gamblers might have reasoned. If they thought of probabilities in that way, they'd expect an increasingly stable frequency of each attribute as the number of trials increased. In the case of a fair roulette wheel, they'd expect black to occur with a frequency of a little under a half. So after lots of black results, aren't red results *owed*, as a result of the law of stability?

No, again, because that would require the system producing the results to have some kind of memory. The wheel would have to somehow 'know' that it was time to give out red results, in virtue of the results it had given before. Or, to put it more precisely, the previous results of the wheel would be exerting causal influence on the future results. But the

TABLE 9.1

Flips	Heads	Tails	Rel. freq. of heads	Tails – Heads
10	0	10	0/100	10
100	40	60	40/100	20
1000	450	550	45/100	100
10000	4800	5200	48/100	200

truth of the law of stability *does not require* any such 'memory' (or causal link between individual results).

There is, though, a reason that it might appear to require this kind of 'memory'. If the relative frequency of red results is to tend towards the relative frequency of black results, as the number of spins increases, then mustn't the number of red results also tend towards the number of black results? The answer is no, although this is not immediately obvious. To see the point, consider the hypothetical results from multiple flips of a coin as shown in table 9.1. The relative frequency of heads results moves closer towards one half (50/100) at each data point, although the difference between the number of tails and heads results *increases* considerably. Hence the law of stability does not entail that runs of one result will be 'matched' by runs of the other, when the process is fair and there are only two possible results.

Thinking in terms of propensities is helpful in order to avoid the gambler's fallacy. In Popper's single case variant, the situation is especially clear. The system (including the process) – e.g. the roulette wheel and operator – has the *same* propensity to generate each result in each particular case. It is a *property* of the system. The past results of the system's operation are totally irrelevant. On Gillies's long run variant, by way of contrast, the system possesses a propensity to generate a specific relative frequency for each result over time. But this is compatible with the results being independent, and does not mean that future results have to 'counterbalance' past results, as illustrated above with reference to the law of stability.

Of course, watching what happens – again, say, on a roulette wheel – *can* give a guide to probabilities, and *thereby* provide some guidance on what will happen in future.

Imagine, for example, that the roulette wheel at Le Grande Casino, mentioned above, was being spun for the first time when the long run of black results happened. If you knew that, as an observer, you might reasonably take the run to be evidence that the wheel was biased. You might think, more particularly, that the probability of black was higher than that of red. This is reasonable. (Possible physical causes include an iron ball being rolled, and strong magnets being hidden under the black compartments.) But it is distinct from committing the gambler's fallacy, because you would be using past data to work out the probability of black on each spin. You would not be assuming that what happened on one spin had any *effect* on what happened on another. Note also the consequence of your hypothetical reasoning. You would be more inclined to bet on black than on red. Those committing the gambler's fallacy did the opposite.

There is also an *inverse gambler's fallacy*, discussed by Ian Hacking (1987), which is worthy of a brief mention. This involves mistakenly inferring what happened in the *past*, rather than what will happen in the future, because of what happens in the present. Or more precisely, in game-like scenarios, it involves the assumption that results at some point in time are *dependent* on earlier results in some way. One of Hacking's examples is as follows:

> [T]hink of a gambler coming into a room, walking to the fair device [for rolling two dice], and seeing it roll double six. A kibitzer asks, 'Do you think this is the first roll of the evening? Or have there been many rolls?' The gambler reasons that since double six occurs seldom, there have probably been many rolls. (1987: 33)

It should be obvious why this is crazy. In fact, *any* result occurs as seldom as any other, provided that the dice are fair. A one followed by a two has the same probability as a two followed by a one, and a one followed by a three, and so on. *Each of the thirty-six indivisible possibilities occurs as seldom as any other.*

Moreover, my own view is that the inverse fallacy need not involve only inferences about *how many* trials have occurred in the past. Consider again the long run of black results in

Le Grande Casino. A gambler who did not know the recent history of results might conclude that there must have been a long run of red results at some point beforehand, perhaps in the near past, which was being 'cancelled out' as he watched. It is interesting, and perhaps somewhat puzzling, that this possibility seems not to have been considered by the gamblers present, when the long run of black results occurred.

2 The Base Rate Fallacy

Imagine that you have applied for life insurance, and that you have had a suite of blood tests as a result. (It's standard for life insurance companies to give blood tests, e.g. for HIV, before deciding whether to offer cover.) One of the blood tests gives a positive result for an unusual disease, which affects one in ten thousand people. The test never gives *false negatives* – if the blood is from a person with the disease, it always gives a positive result. But the test sometimes gives *false positives* – it sometimes gives a positive result when the blood is *not* from an infected person. To be more specific, it gives false positives 1 per cent of the time. What's the probability that you have the disease?

Are you tempted to say 99 per cent? Most people would be, on the evidence of psychological experiments. So they would take the test result to show that they had the unusual disease. But this is a mistake, because it ignores a crucial piece of available information, namely the *base rate* of the people with the unusual disease. Only *one in ten thousand* suffer from it, recall. Yet the results of the test, if run on everyone, would (probably) lead us to think that many more than one in ten thousand did.

In this case, thinking of all the information in a way consistent with the relative frequency interpretation of probability not only helps to avoid making the mistake, but also makes it easy to see the correct answer. Let's now try this:

1 1 in 10,000 has the disease.
2 Everyone with the disease tests positive for it.
3 Therefore, 1 out of every 10,000 tests positive and has the disease.

4 9999 out of 10,000 do not have the disease.
5 99.99 out of every 9,999 test positive but do not have the disease.
6 Therefore, 99.9 out of every 10,000 test positive and do not have the disease.

So what's the relative frequency of people who have a positive result and *really* have the disease? It's simple to work out with a few more lines of reasoning:

7 100.9 out of every 10,000 test positive for the disease. (Addition from 3 and 6.)
8 1 in every 100.9 people testing positive has the disease. (From 1 and 7.)
9 Therefore, the probability of having the disease when testing positive is $1/100.9 \approx 1/101$.

So you shouldn't panic! (Since we're talking in a relative frequency way, you shouldn't, strictly speaking, assign a probability to yourself. You can think of yourself as a person picked at random from the collective, though.) If the insurance company refused to insure you on the basis of the positive result, they would be rather unfair. The risk is minimal. Alas, insurance companies are often unfair, in so far as they err too strongly on the side of caution. That's a problem we can't solve here. I wish we could.

It is worth noting that Bayes's theorem, which is discussed in Appendix B, is of great help in avoiding this fallacy. Consider again the example above. Let h be 'You have the disease'. And let e be 'You have tested positive for the disease'. We are interested in $P(h, e)$, or the *posterior probability* of h, which we can calculate using the theorem. If you would like to have a go at showing this, consult the worked example in the appendix. Here's the information given in the example above, expressed formally, as a guide:

1 $P(e, h)$ is 1; if you've got the disease, the test will definitely show it.
2 $P(e, \neg h)$ is 1%, or 0.01; the test has a small chance of giving a 'false positive'.
3 $P(h)$ is 1/10000, or 0.0001; the base rate of the disease is very low.

3 The Inverse Fallacy

The so-called 'inverse fallacy' involves confusing a conditional probability for its inverse; or to put it formally, it occurs when $P(p, q)$ is confused with $P(q, p)$. Why this is wrong – if it is not immediately obvious – is best illustrated by thinking in terms of the logical interpretation of probability, and cases of logical entailment. Let p be 'All rabbits are black and Tim is a rabbit' and q be 'Tim is black'. $P(q, p)$ is equal to one, because p entails q. But q does not entail p. For all that q says, Tim could be a man, a snake, a dog, a cat, a horse, or even a toy. So q hardly indicates that 'Tim is a rabbit'. Moreover, the mere fact that there is a black thing (which happens to be called 'Tim') does not confirm that 'All rabbits are black'.

You might think this mistake is *so* obvious that only fools would make it. But remarkably, there is evidence that this mistake *is* made regularly. What's worse, the mistake is even made by (widely acknowledged) experts, in situations where the consequences are serious. Jonathan Koehler (1996) discusses how this happens in trials, when forensic scientists are questioned about 'matching' genetic material to people. For example, imagine there is a match between my DNA, and that of a hair found at the scene of the crime. A forensic scientist might state the probability of a match occurring given that the hair is not mine: $P(match, \neg mine)$. We often hear of such testimony. It could go like this: 'The chance of there being a DNA match, if the hair is not Darrell's, is less than one in a million.' But this is completely different from $P(\neg mine, match)$. So it would be completely wrong to conclude: 'The chance of the hair not being Darrell's, given the match we've found, is less than one in a million.' Yet some forensic scientists conclude, or at least *say*, precisely that. Here's Koehler's (1996, n. 8) example, which uses a court record:

> After testifying that a DNA match was found between blood from a murder victim and blood recovered from a blanket, an FBI scientist in a Florida case was questioned by a prosecuting attorney as follows:

> Q: And in your profession and in the scientific field when you say match what do you mean?
> A: They are identical.
> Q: So the blood on the blanket can you say that it came from Sayeh Rivazfar [the victim]?
> A: With great certainty I can say that those two DNA samples match and they are identical. And with population statistics we can derive a probability of it being anyone other than that victim.
> Q: What is that probability in this case?
> A: In this case that probability is that it is one in 7 million chances that it could be anyone other than the victim.

Frightening, isn't it? What's worse, experiments by Koehler (1996) suggest that such misleading evidence can sometimes strongly influence jurors to give guilty verdicts. No surprise there! It comes as some comfort, therefore, that thinking in logical terms – such as those above, used in presenting the fallacy – can help to avoid the error. This is a finding of Pawel Kalinowski, Fiona Fidler, and Geoff Cumming (2008). They also found that encouraging students to use Bayes's theorem, covered in Appendix B, reduced their susceptibility to committing the fallacy. This is not so surprising, since both strategies encourage students to represent statements as symbols – propositional variables, like p and q – and consider unidirectional relationships between those symbols.

In closing, it's worth adding that the inverse fallacy may also account for why *some* people tend to ignore base rates. They may think they've already been given the information they seek, e.g. $P(q, p)$, when they've been given rather different information, e.g. $P(p, q)$.

4 The Conjunction Fallacy

Consider the following passage and question:

> Frank is 35 years old, tall, and athletic. He excelled at many sports in his youth. He had a special gift at football from an early age, and played in central midfield for

> England's under-18 team. He went on to play for a national club. He retired from professional football at age 29.
>
> Which is more probable?
>
> (1) Frank is a teacher.
> (2) Frank is a PE teacher, and is coach of his school's football team.

Did you get the right answer? The passage above is based on one used in an experiment by Amos Tversky and Daniel Kahneman (1982), which is as follows:

> Linda is 31 years old, single, outspoken, and very bright. She majored in philosophy. As a student, she was deeply concerned with issues of discrimination and social justice, and also participated in anti-nuclear demonstrations.
>
> Which is more probable?
>
> (1) Linda is a bank teller.
> (2) Linda is a bank teller and is active in the feminist movement.

Most of the subjects in Tversky and Kahneman's experiment selected the second option. But that's a mistake. The reason is simple. The probability of a conjunction – of events or propositions – cannot be higher than that of one of the two conjuncts. Or to put it formally: $P(p \,\&\, q) \leq P(p)$ and $P(p \,\&\, q) \leq P(q)$. (Making the mistake in response to my example is rather worse. It involves not only 'Frank is a teacher' (p), but also 'Frank teaches PE' (q) and 'Frank is coach of his school's football team' (r). And it should be *more* obvious that $P(p \,\&\, q \,\&\, r) \leq P(p)$ than it is that $P(p \,\&\, q) \leq P(p)$.)

Tversky and Kahneman's original experiments were on students. Incredibly and frighteningly, though, they later found that the same kind of mistake was made – in 91 per cent of cases – by physicians who were asked to diagnose patients based on descriptions of their symptoms. The physicians thought the patients were more likely to suffer from two problems, rather than one. For the details, see Tversky and Kahneman (1983: 301–2).

So why does the mistake occur? A simple answer is that there's no information in the passages concerning banks, bank tellers, teachers, etc. Hence, people are drawn to the options that seem most relevant to the material discussed. There is some dispute about exactly why this is, but we won't go into it here. Reading Tversky and Kahneman (1982) is a good way to start, if you're interested.

Many different interpretations of probability are helpful in illustrating this fallacy. Thinking about gambling scenarios and rational betting quotients, as discussed in Chapter 4, is one option. A rational person (facing a smart bookie) would not have a higher betting quotient on two horses winning separate races than on just one of those horses winning its race.

Interestingly, studies have shown that the conjunction fallacy is less likely to be committed – in some cases, at least – when the question is posed in a frequency-based way, with mention of a specific *reference class*. For example, Klaus Fiedler (1988) found that the percentage of people committing the fallacy dropped significantly when he posed the 'Linda question' as follows:

> There are 100 persons who fit the description above (i.e. Linda's). How many of them are:
>
> (a) bank tellers
> (b) bank tellers and active in the feminist movement

This concludes our coverage of named fallacies. We have seen that thinking in terms of specific interpretations of probability can help to illustrate, and often avoid, committing them.

5 The Monty Hall Paradox

I include this final section partly for a bit of fun, although there are some morals to be drawn from it. Not only am I going to present a rather fascinating paradox – at least, it is often called a 'paradox' – but also an amusing story about how numerous arrogant male professors got egg on their faces after disagreeing with an amateur woman (who quit university part way through a philosophy degree).

The story concerns a puzzle posed in Marilyn vos Savant's column in *Parade* magazine, 'Ask Marilyn', in the early 1990s. Marilyn vos Savant is best known for her high scores on IQ tests; for five years, she was listed in the *Guinness Book of World Records* under 'Highest IQ'. (Incidentally, this category was discontinued because of the unreliability of IQ tests. Moreover, vos Savant's childhood test was interpreted incorrectly as showing that she had an IQ of 228, when the maximum possible rating is '170+' according to the test manual. Incidentally, I had a childhood IQ test, at around 7 years of age, where I also scored 170+. Sadly, I did not rise to fame and fortune. It was quite a nice result, nonetheless, because I'd been sent to a psychiatrist at my school's request. On account of my misbehaviour, they thought that I had learning difficulties. I suspect that they had *teaching* difficulties!)

The puzzle posed – in fact, this is just a different version of a puzzle presented by our old friend Bertrand (1888), now known as his *box paradox*, which was already known to some as 'the Monty Hall problem' (see Selvin 1975) – is as follows (see vos Savant 2014):

> Suppose you're on a game show, and you're given the choice of three doors. Behind one door is a car, behind the others, goats. You pick a door, say #1, and the host, who knows what's behind the doors, opens another door, say #3, which has a goat. He says to you, 'Do you want to pick door #2?' Is it to your advantage to switch your choice of doors?
>
> *Craig F. Whitaker*
> *Columbia, Maryland*

And here's how vos Savant (2014) responded:

> Yes; you should switch. The first door has a 1/3 chance of winning, but the second door has a 2/3 chance. Here's a good way to visualize what happened. Suppose there are a million doors, and you pick door #1. Then the host, who knows what's behind the doors and will always avoid the one with the prize, opens them all except door #777,777. You'd switch to that door pretty fast, wouldn't you?

Is vos Savant right? Many academics, including mathematics professors, said 'No!' Moreover, they continued to do so after she defended her answer in greater detail. Here's an entertaining selection of excerpts from these academics' letters, many of which are reproduced by vos Savant (2014):

> Whether you change your selection or not, the odds are the same. There is enough mathematical illiteracy in this country, and we don't need the world's highest IQ propagating more. Shame!
>
> You blew it!...As a professional mathematician, I'm very concerned with the general public's lack of mathematical skills. Please help by confessing your error and in the future being more careful.
>
> May I suggest that you obtain and refer to a standard textbook on probability before you try to answer a question of this type again?
>
> [Y]ou will receive many letters on this topic from high school and college students. Perhaps you should keep a few addresses for help with future columns.
>
> You made a mistake, but look at the positive side. If all those PhD's were wrong, the country would be in some very serious trouble.

Wow! It's hilarious that the country was indeed in 'some very serious trouble', if the final sentence above were true, since all those PhD's *were* wrong (given some of the assumptions they shared with vos Savant, which we'll bring out below). To see why, it is helpful to think in terms of long run propensities, in repeats of the game show.

Let's think carefully about the process, and list results, as we go along.

> Step One: The car is randomly placed behind one of three doors.
>
> (This assumption was not made explicit. It is possible for others to take its place without changing the result, e.g. for the car to be placed intentionally behind the same door

– or a particular sequence of doors – on repeats but for you to choose a door at random in step two.)

Step Two: The contestant selects a door.

Result One: The door selected in step two will have the car behind it approximately one-third of the time, in the long run (and exactly one-third of the time, in the limit).

Note, at this point, the importance of the randomness assumption made in step one (which is not clear in the original question or in the assumptions listed in von Savant's reply). Without this assumption it might be true that the contestant always selects door one, and the car is always behind door one, for example. Let's now think through the remainder of the process.

Step Three: If the door the contestant selected has the car behind it, then the host must open one of the other two doors at random. If the door the contestant selected does not have the car behind it, then the host must open the unselected door that does not have the car behind it. Hence, the host opens an unselected door that does not have the car behind it *in a way that does not signal where the car is*.

(This is a further crucial assumption missing from the original question and vos Savant's reply. However, vos Savant *does* stipulate that the host 'always avoids the one [door] with the prize'.)

Result Two: The host's action in step three *in no way alters* result one, namely that the door selected in step two will have the car behind it only one-third of the time. From this – and the fact that the car is behind one of the two unopened doors remaining after step three – it follows that the unopened door *not selected in step two* will have the car behind it two-thirds of the time.

Step Four: The contestant is given the choice to change their selection of door (to the unopened door *not selected in step two*).

Step Five: The contestant wins whatever is behind the door they choose.

Result Three: Changing door selection will win the car two-thirds of the time.

If you still don't buy this, you can do the experiment. Or you can find a computer simulation online, if you Google 'Monty Hall simulator'. (At the time I write, there's one at: http://stayorswitch.com). But it is crucial to understand the role of the further assumptions, mentioned after step three, which do *not* appear in the question. To illustrate, imagine that step three is different. Imagine that the host *must* open one of the unselected doors, but that he doesn't know where the car is, so the door he opens *can* have the car behind it. If the door he opens *does* have the car behind it, the game ends (with the contestant winning only a goat) before step four is reached. The long run propensities for changing door selection at step four *are now different*. Think of it this way. One-third of the time, the contestant will pick the correct door at the start. Another third of the time, the host will pick the correct door in (the modified) step three. (If this is not immediately obvious, think as follows. Two-thirds of the time, the contestant does not pick the correct door. And the host will open the door with the car behind it only on *half* of those occasions. Thus, the host will find the car one-third of the time overall; two-thirds multiplied by one-half is one-third.) In the final third of the time, the game will reach step four and the car will be behind the unopened door that the contestant *did not* select initially. Overall, then, step four is only reached two-thirds of the time. Half of the time, at that point, the car is behind the door selected at step two. The other half of the time, it's behind the other unopened door. Therefore, there is no advantage (or disadvantage) to changing door preference at step four.

There are two morals to draw from the story. The first, which is also supported by other findings we've discussed in previous sections of this chapter, is that simple-looking probability problems can be tricky to solve without proceeding carefully and systematically. And thinking in terms of different interpretations can help you to better understand such problems, in many cases.

The second moral is as follows. It's important to make sure that a probability problem is well defined, before attempting to tackle it. And thinking in terms of different interpretations can sometimes help to illustrate whether a problem is well defined. On this occasion, thinking in terms of long run propensities proves helpful *because it focuses attention on the process or the system*. (Thinking in terms of degrees of belief in the outcomes, on the other hand, doesn't necessarily do that.)

Further Reading

Many references have already been provided to psychological work on the fallacies covered above. A fun introductory place to look for details of more similar errors in reasoning – as well as more coverage of some of the fallacies we've discussed – is Reinhart (2015). See also the website: www.statisticsdonewrong.com.

10
Probability in the Humanities, Natural Sciences, and Social Sciences

And now, the end is near. And as we face the final curtain, we will take a look at how the interpretation of probability makes a difference to how we can understand (and sometimes apply) a selection of theories from the humanities, natural sciences, and social sciences. We'll look at four specific areas, in philosophy, biology, economics, and physics: confirmation theory, Mendelian genetics, game theory, and quantum theory.

1 Confirmation Theory

Philosophers of science have long been concerned with what it means to say that a theory is confirmed, or disconfirmed, by evidence. After all, most people, scientists included, think that the central theories of contemporary science are well confirmed by all the available evidence, even if they recognize that they're not entailed by such evidence. Understanding how confirmation occurs may also have some consequences for studies of scientific method. One interesting debate, for example, concerns whether theories can be confirmed merely if they *accommodate* known evidence, rather than make *novel predictions*.

Now it is natural to think that confirmation comes in degrees, and can be measured on a numerical scale, because different competing theories can be confirmed to different extents by the same evidence. For example, T_1 might be confirmed more than T_2 by e, and T_3 might be confirmed more than T_1 or T_2 by e. So if we represent the confirmation of a theory T given evidence e as $C(T, e)$, we may say that $C(T_3, e) > C(T_1, e) > C(T_2, e)$ under such circumstances. Here's a simple example. Imagine that the theories are about the relative frequency of heads results in 10,000 coin flips. Let e be 'In the first 100 flips, the relative frequency of heads was 0.55.' Let T_3 be 'The relative frequency of heads results will be between 0.45 and 0.55 overall', T_1 be 'The relative frequency of heads results will be between 0.49 and 0.51 overall', and T_2 be 'The relative frequency of heads results will be between 0.7 and 0.72 overall.' Evidently, $C(T_3, e) > C(T_1, e) > C(T_2, e)$.

It is also natural to say things like 'Einstein's theory of general relativity is probably true', which is a casual way of saying 'Einstein's theory of general relativity is probably true *relative to the available evidence*', as well as 'Einstein's theory of general relativity is highly confirmed'. And perhaps this is no coincidence. Should we say that 'T is highly probable (on e)' is *equivalent* to 'T is highly confirmed (by e)'? A popular answer to this question, but by no means the only answer, is 'Yes'; $C(T, e) = P(T, e)$. This is also a rather convenient view, because Bayes's theorem, which is discussed in Appendix B with a worked example, may then be used to calculate confirmation values.

If the confirmation of a theory is equivalent to its probability, then it is obvious that the interpretation of probability matters in how we understand confirmation. (We will come to *how it matters* in short order.) But many who think that confirmation is not *the same as* probability – i.e. hold that $C(T, e) \neq P(T, e)$ – nonetheless think that confirmation values should be *defined in terms of probabilities*. One view, for example, is that how much some evidence (e) confirms a theory (T) can be measured by considering to what extent the theory 'predicts' the evidence: $C(T, e) = P(e, T) - P(e)$. (This can have a negative value, hence *isn't* a probability.) So how we interpret probability, when it comes to scientific theories, matters for many (possible and actual) accounts of confirmation.

How, exactly, does it matter? One crucial way is with respect to the *objectivity* of confirmation in science. If the probabilities involved are purely subjective, for instance, then there is no *objective* answer to how confirmed any scientific theory is, at any point in time (unless, perhaps, the theory is *inconsistent* with the evidence); the confirmation value may differ considerably, from rational person to rational person. So saying 'T is highly confirmed by e' should prompt the question 'For whom?' And there is no obvious reason to take the view of a scientist as any better than that of a priest, when each accepts the same evidence, provided the degrees of belief of each satisfy the axioms of probability (and some other minor constraints – e.g. respect entailment relations – which were mentioned in Chapter 4). Something similar is true if the probabilities are understood as purely intersubjective; the confirmation value may rationally differ, from group to group. For example, the Catholic Church and the Royal Astronomical Society might rationally prefer different astronomical theories while accepting the same evidence (e.g. observation statements made using telescopes).

On a logical view of probability (and in some cases, an objective Bayesian view), by way of contrast, the relationship between the evidence and the theory is *unique and fixed*. So if the Catholic Church and the Royal Astronomical Society disagree about which astronomical theory is more confirmed by the available evidence, then at least one of those groups *must* be wrong.

To bring out the difference more precisely, let's consider a historical scenario using the measure $C(T, e) = P(e, T) - P(T)$, which was mentioned above. This concerns the so-called 'Poisson bright spot'. In 1819, the French Academy of Sciences announced that their grand prize for the year would be awarded to the best paper they received on the diffraction of light. (Diffraction is explained, from a modern point of view, in the subsequent section on quantum mechanics; the details need not concern us at the moment.) Augustin Fresnel was one entrant to the competition, and developed a *wave* theory of light in the paper that he submitted. However, several members of the judging panel instead preferred a *corpuscular*

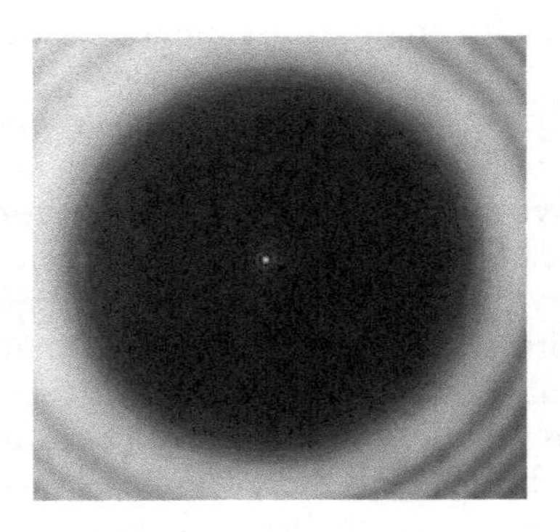

FIGURE 10.1 The Poisson bright spot
Reproduced with permission from *The American Journal of Physics* 44: 70. Copyright P. M. Rinard, 1976, American Association of Physics Teachers.

theory of light, in which light is composed of particles. One such judge, Simeon Poisson, showed that a bright spot would appear in the centre of the shadow cast when an opaque disc is illuminated, if Fresnel's theory were correct. And he took himself, and his fellow judges, to know that no such bright spot appears. That is, on the basis of everyday experience with shadows. (Artists also drew shadows, by geometrical projection, as solid patches.) So he argued that Fresnel's theory was false. Luckily for Fresnel, however, the chair of the judging panel, namely François Arago, insisted on performing the experiment described by Poisson. And the bright spot was found! See figure 10.1.

Fresnel won the grand prize! And we may understand this, in terms of our equation for C(T, e), as follows. The existence of the Poisson bright spot (e) was significant because it was a consequence of Fresnel's wave theory of light (T) (in conjunction with other uncontroversial claims), but was '*unexpected*' beforehand. P(e, T) = 1, whereas P(e) was extremely low, so C(T, e) was very high. (I mention 'other uncontroversial claims'. If desired, we can explicitly mention background information (b), which includes old evidence, in our confirmation formula; i.e. use C(T, e, b) = P(e, T & b) – P(e, b). I leave b out just to keep the formulae simple.)

What should we make of the use of '*unexpected*' above, i.e. of P(*e*) being extremely low before the experiment showing the existence of the bright spot was performed? Here's where the interpretation of probability matters. Were the judges just *surprised* by the prediction because they *happened* not to have expected it? (Were the personal probabilities simply low for the members of that group?) Then the force of the experiment appears to have been purely psychological. On the other hand, if the only rational degree of belief in P(*e*) was extremely low, given the information available before the experiment – that is, on a logical or objective Bayesian view of probability – then the experiment appears to have had some *objective* force. It revealed a phenomenon that it was *irrational to believe in (on the accepted evidence) beforehand*. And Fresnel's theory predicted the phenomenon.

So in summary, how we treat probabilities in the context of confirmation theory can make a big difference to how we understand the epistemic status of well-confirmed contemporary scientific theories.

2 Mendelian Genetics

In the nineteenth century, Gregor Mendel performed numerous breeding experiments with pea plants (*pisum sativum*), and sowed the seeds – yes, the terrible pun is intended! – for modern genetics. Specifically, he was interested in how different *traits* are inherited. If a plant with yellow peas is crossbred with a plant with green peas, for instance, then what colour will the peas of the resulting plant be? And what if a tall pea plant is crossbred with a dwarf pea plant? How tall will the resulting plant be?

Mendel found that there are patterns in how such traits are inherited, and that they do not simply 'blend' (as many had thought). For example, when one parent pea plant has purple flowers and the other has white flowers, the flowers of the offspring plants are *not* pink. They are either white or purple, as we can see by considering Mendel's results in greater detail.

Mendel began by cross-pollinating white-flowered plants with purple-flowered plants. Only purple-flowered plants

	X	X
x	Xx	Xx
x	Xx	Xx

FIGURE 10.2 Punnett square for parent cross-pollination

resulted (in the first generation, or G1). However, he then allowed the G1 plants to self-pollinate, and produce a further generation of plants (G2). He found that approximately one-quarter of the plants in G2 were white, whereas the other three-quarters were purple. (His specific results for G2 were as follows: 705 plants had purple flowers, and 224 plants had white flowers. So the probability of a G2 plant being white appeared to be one quarter, and the probability of a G2 plant being purple appeared to be three-quarters.)

Mendel explained this by positing that each pea plant contains a *pair* of genes determining its colour. He also posited that only two *types* of colour genes (or two *alleles*) were present in the pea plants. Let's call these X (for purple) and x (for white). Mendel's idea was that the plants he started with, which he cross-pollinated, contained the pairs XX and xx. The resultant plants would take one gene from each parent, so could only end up as Xx (or xX, which is the same combination; the order does not matter). We can represent this in a diagram known as a Punnett square (see figure 10.2). This is named after the British geneticist Reginald Punnett. (I recently discovered that I went to the same school as he did, albeit many decades later. What's the probability of that?) The possible *alleles* from the sperm are shown on the side – here, from the xx (white-flowered) plant – whereas the possible alleles from the egg – here, from the XX (purple-flowered) plant – are on the top.

	X(P = ½)	x(P = ½)
X(P = ½)	XX(P = ¼)	Xx(P = ¼)
x(P = ½)	Xx(P = ¼)	xx(P = ¼)

FIGURE 10.3 Punnett square for G1 self-pollination

So provided that the X allele is *dominant* over the x allele, all the offspring plants (in G1) will be purple-flowered. And Mendel took his experiments to indicate this dominance of X over x. (The norm is to use capital letters to represent dominant alleles, and lower case letters to represent recessive alleles.)

Now when the plants in G1 reproduce (by self-pollination), the possibility of having XX offspring and xx offspring is present, as shown in figure 10.3. More precisely, the sperm and egg might both contain X alleles, or both contain x alleles. Hence, white-flowered (xx) offspring reappear.

You will see that probabilities are marked on figure 10.3. The idea is that the probability that the sperm contains X equals the probability that the sperm contains x, and that the sperm can only contain X or x. Ditto when 'egg' replaces 'sperm' in the previous sentence. Thus the probabilities for each internal square – which, as you will see, are determined by multiplying the probabilities for the relevant row (sperm allele) and column (egg allele) – are also equal. As Xx appears twice, its probability overall is one half; the probabilities of XX and xx are each equal to one quarter.

Let's now think about how to interpret these probabilities. And this is an interesting case, because I think that there is a single correct interpretation. My argument is as follows. First, the use of a world-based probability is appropriate, because we are dealing with real frequencies in the world, the existence of which we want to *explain*. Think of it this way. (Assume there are two types of alleles responsible for colour in the original plants, and that pairs of such alleles exist in each pea plant under discussion, as discussed above.) *It could instead have been true that sperm was much more likely to carry x than X, and also that eggs were much more likely to carry x than X*. That's to say, x might have been 'easier' to carry than X for some reason. And then the frequency of xx (white-flowered) offspring at the G1 self-pollination stage would (probably, if the law of stability holds) have been considerably higher in Mendel's experiment. Thus the frequency measured is an interesting empirical fact.

Second, let's think about which world-based interpretation is suitable. Are we to say that the probabilities are *just* the actual frequencies? No, because the actual frequencies differ slightly from the numbers we want (e.g. 224/705 is not 1/3). So are we to say instead that they are hypothetical frequencies? Only on pain of accepting that there are no propensities to *explain* why those hypothetical frequencies take those values. (Admittedly, one *could* accept that propensities exist, but insist that these should not be called 'probabilities'. But we won't take this possibility seriously here, as it would be a big diversion.) So we should check out the propensity options before reverting to the relative frequency one.

Third, let's consider which version of the propensity view might do the trick. Are *single case propensities* present in the situation covered by figure 10.3, for example? It seems not. Think of an individual egg meeting an individual sperm. There is a fact of the matter about whether the egg carries the x or X allele, and the same is true for the sperm. So there is a fact of the matter about what the flower colour of the resulting plant will be, in any single case of cross-pollination.

Hence, the long run propensity view seems to be the right one for the Punnett square. There are (approximately) as many eggs of any one (possible) type as of any other, just as there are (approximately) as many sperm of any one (possible) type as of any other. And there's no barrier to sperm of any particular type interacting with eggs of any particular type. So there's nothing about the system to favour any specific kind of interaction; in the long run, any possible kind of interaction should be expected to occur (approximately) as often as any other.

Let's again use self-pollination in G1 (depicted in figure 10.3) as an example. There are (roughly) as many *X*-carrying eggs as *x*-carrying eggs (over time), and (roughly) as many *X*-carrying sperm as *x*-carrying sperm (over time); moreover, there are many eggs and sperm, with many opportunities to interact. Any sperm is free to interact with any egg; there is nothing about the system to favour eggs of either type interacting with sperm of either type. Thus it is reasonable to expect that *X*-carrying eggs will interact with *x*-carrying sperm (approximately) as often as they interact with *X*-carrying sperm, and so forth, in the long run.

You may have thought even deeper. You may have noticed that I didn't consider *how* or *why* there turn out to be (approximately) as many sperm (or eggs) of each possible type (over time). So might there not be single case propensities at play *at this more fundamental level*? To answer this, we would need to think about how sperm and eggs are produced, and that would take us into much more complex territory. It seems to me, however, that modern biology takes this to be a deterministic process; so I think it is reasonable to believe that the frequencies of eggs and sperm of each type are also due to long run propensities.

3 Game Theory

'Game theory' is exactly what it sounds like – theory about how to play games successfully. It is more important than may at first be apparent, because many situations in daily life are like games (or are suitably game-like to be covered by the theory). For example, game theory covers the case of two car drivers meeting each other on a narrow country lane, where one has to reverse for some distance in order to let the other past. Think about the situation. If both drivers refuse to reverse, then neither gets to where she wants to be on time. But whichever driver reverses to let the other by will get to her destination more slowly than she would if the other reversed, because she will lose ground. Hence, we can think of this as a game involving the two drivers.

So how do probabilities matter in game theory? Let's begin by considering a two-player example of a much-discussed

		Player Two	
		Cooperate	*Defect*
Player One	*Cooperate*	–1, –1	–3, 0
	Defect	0, –3	–2, –2

FIGURE 10.4 A symmetrical two-player prisoner's dilemma

game, namely the prisoner's dilemma. This involves two people who are arrested, after committing a criminal act together, and undergo individual questioning by the police. Neither can communicate with the other, and neither cares what happens to the other. Each is given a choice. She can *defect* by implicating her partner in crime, or she can *cooperate* by refusing to implicate her partner. She knows that if both *cooperate*, then they will each receive one year in prison. She knows that if both *defect*, then they will each receive two years in prison. And she also knows that if one *defects* but the other *cooperates*, then the former will go free, whereas the latter will receive three years in prison. Finally, each prisoner knows the game will not be repeated, and that there will be no opportunity for retaliation from the other.

The outcomes are depicted in figure 10.4. Zero represents no prison time, minus one represents one year of prison time, and so forth; and we assume that two years is twice as bad as one year, etc. (It's possible to substitute other punishments, to make sure this kind of scaling is correct for the participants, if desired.) The result in the bottom right corner, for example, represents two years in prison for player one, and two years in prison for player two. The players' situations are *symmetrical*, in so far as we could swap the 'Player One' and 'Player Two' labels and the diagram would still provide an accurate representation of the outcomes (penalties) given the choices.

Now it is possible to use probabilities to calculate how to *minimize the expected sentence*, as either player, by assigning probabilities for the other player to *cooperate* or *defect*. So let player two have a probability of n of *cooperating*, and thus a probability of $1 - n$ of *defecting*. Player one's expected sentence is $-1(n) + -3(1 - n) = 2n - 3$ for *cooperating*, and $0(n) + -2(1 - n) = 2n - 2$ for *defecting*. It turns out in this

		Player Two	
		Stag	*Hare*
Player One	*Stag*	2, 2	0, 1
	Hare	1, 0	1, 1

FIGURE 10.5 A symmetrical stag hunt

case that the probability value is irrelevant; the expected sentence for *cooperating* is always one year longer than that for *defecting*. (You can also see this by stepwise reasoning. Imagine player two *cooperates*. Then it's best for player one to *defect*. Imagine player two *defects*. Then it's best for player one to *defect*. As player two can only *cooperate* or *defect*, it's therefore always best for player two to *defect*.)

However, there are similar games where the values of probabilities *are* relevant. Consider now a different symmetrical game, called the Stag Hunt, an example of which is shown in figure 10.5. The idea behind this game is as follows. (Although note, as intimated earlier, that many other situations may effectively be represented as 'stag hunts'.) Both players are hunters, and have to select whether to hunt a stag or catch a hare (in order to provide food and other assets, e.g. skins, for themselves and their families). It takes both players to catch a stag, which they will do if they cooperate with one another (i.e. both select *stag*, which you can think of, by analogy with the prisoner's dilemma above, as a *cooperate* option). But a player who elects not to cooperate is guaranteed to catch a hare. Thus if one player goes stag hunting while the other instead goes to catch a hare, then she will end up with nothing (although the other will get a hare). A stag is twice as valuable as a hare, for each hunter, if it is caught. (Each number in figure 10.5 may be said to represent a *utility*, or the ultimate satisfaction that can be derived from the catch. *Utility* is an important concept in economics and game theory, although we cannot discuss it in depth here. If you prefer, you can instead imagine the hunters will always sell the carcasses of whichever animals they kill, and think of the numbers as representing units of monetary income, such as tens of US dollars. You can then replace the subsequent talk of 'expected utility' with 'expected income'.)

Now let's do a probability calculation, from the perspective of player one, similar to the one we did for the prisoner's dilemma scenario above. (Again, we assume that player one is only interested in maximizing her own profit.) Let player two have a probability of n for joining the hunt for a *stag*, and a probability of $1 - n$ for individually catching a *hare*. Player one's expected *utility* is $2n + 0(1 - n) = 2n$ for choosing *stag*, and $1n + 1(1 - n) = 1$ for choosing *hare*. So if $2n > 1$ – i.e. $n > 0.5$ – then choosing *stag* is the better option for player one. But if $n < 0.5$, then it is better for player one to choose *hare*.

So now let's think about how to interpret the probabilities. In fact, this is a situation where either information-based or world-based views can be used, because we can think either of (a) what's rational for a player to do given her own expectations (and/or her available relevant information about a scenario), or (b) what's actually best for a player to do given the situation in which she finds herself.

Let's begin by thinking in the first way, (a), and using the subjective interpretation of probability to illustrate. Let's say player one thinks that the probability of player two choosing stag, n, is just 0.4. (And she also thinks the probability of player two choosing *hare* is 0.6, etc. All her game-related degrees of belief satisfy the axioms of probability.) If player one nonetheless opts for *stag* – while being interested only in maximizing her own profit in the situation (or 'game') – then she has acted *irrationally*. Indeed, we can think of game theory as showing just that – as showing how she ought to behave, given her values and personal probabilities.

But we can also think in the second way, (b); and let's use a relative frequency in the limit view (ignoring some of the problems there may be with this) in order to get a grip on how. Now we don't care about what player one thinks; we are just interested in what strategy will cause her to win most *in an infinite series of repeats of the game*. The value for n tells us the relative frequency, in the limit, with which player two will choose to *stag*. If this is less than 0.5 – and this is *the only information we have about how player two will bet* – then the best strategy for player one is to select *hare* each time.

Sometimes when we think about a series of games, we also want to take into account that the results in each repeat *will not be independent*. Imagine, for example, that player one chooses *hare* in the first game, whereas player two chooses *stag*. In the second game, as a result, player two might want to 'punish' player one by selecting *hare*. On the other hand, player two might be inclined to select *stag* in any given game provided that she has seen player one select *stag* repeatedly in the past (at that point). More complex considerations such as these can also be taken into account while thinking in terms of world-based probabilities. For example, we might consider relative frequencies in the limit of player two's responses in particular *subsets* of games, e.g. games where player one has selected *hare* on the previous iteration. But we will not explore such possibilities any further here.

4 Quantum Theory

Don't be scared, even if you know almost no physics. This won't be as hard as you expect! I'm not going to present quantum theory in any depth here. It's interesting, though, because probabilities are *built into it*. In other words, probabilities are a fundamental part of the mathematics of the theory. But how that mathematics should be *interpreted* is a different matter. There is still much disagreement about this, among physicists and philosophers of physics.

To focus our discussion, I will use a single example. It concerns *diffraction*, which happens to waves when they encounter obstacles. Waves interfere with one another. It's easy to see this by playing in the bath. Drop one object in, and you will see a beautiful circular surface wave appear. Drop two similar objects in simultaneously, some distance apart, and you will create two waves that will meet. At some points, the interference will be *destructive* – that's to say, the waves will cancel one another out either partly or wholly. (The two waves will cancel one another out completely when one wave is at a peak, and the other is at a trough, provided that the peak height is the same as the trough depth.) At other

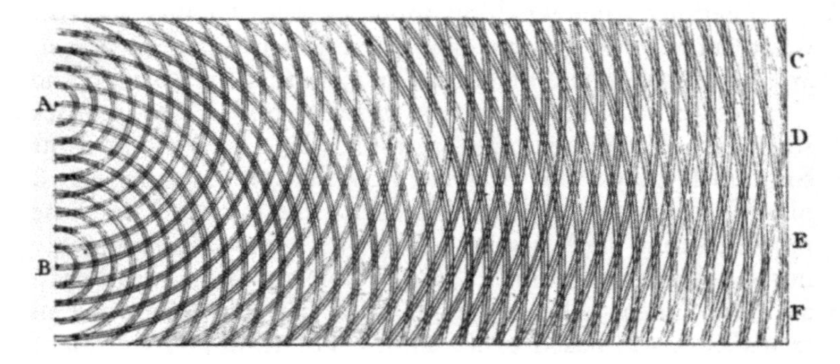

FIGURE 10.6 Diagram of Young's interference experiment [Wikimedia Commons/Quatar]

points, the interference will be *constructive* – that's to say, the intensity will be increased. (If two peaks meet, then a single higher peak will result. If two troughs meet, then a single deeper trough will result.)

Now one of the classic experiments in physics, which was originally performed by Thomas Young in 1803, involves passing light from a single source through two thin slits. This results in an interference pattern, which Young took to support the hypothesis that light is a wave. (In the time that Young was writing, scientists were arguing about whether light is a wave or a particle; he was around at the same time as Poisson, Fresnel, and Arago, whom we mentioned in the earlier section on confirmation theory. Nowadays, however, it's normal to think of light as a *wave particle*, for reasons we'll touch on shortly. You don't have to take this literally. Rather, you may accept only that light *behaves* like a wave in some situations, and *behaves* like a particle in others.) Young's own diagram of the experiment is shown in figure 10.6.

The sketch in figure 10.6 gives a good idea of how the waves meet. The sort of pattern that results, in an actual experiment, is shown in figure 10.7. It's possible to create the same kind of pattern with circular water waves, like the ones I mentioned previously. Try tapping down two similar objects rhythmically, on the surface of some still water. You will see a standing pattern emerge, if you do it properly. A large, well-illuminated, body of water is best.

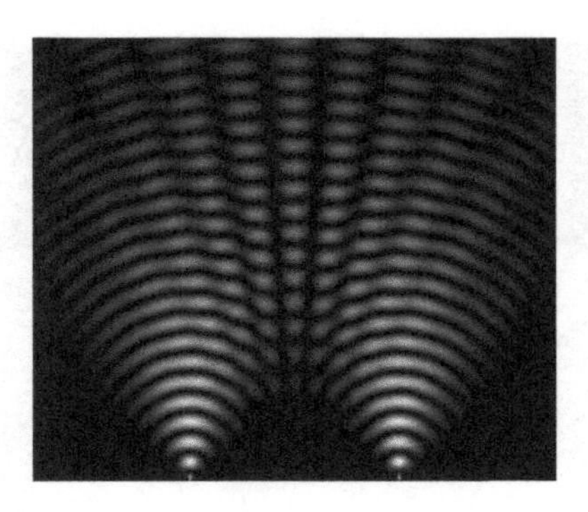

FIGURE 10.7 Young's interference experiment [Wikimedia Commons/Ffred]

Now with light, it's possible to use a screen to view the effects of diffraction on a plane parallel to the one on which the slits are. The intensity of the light on such a screen varies as shown below, in figure 10.8. Part (a) shows how the image on the screen appears; the light is brightest in the middle, and becomes less bright as one moves outward. Part (b) illustrates the intensity of light graphically, and makes it easier to appreciate how it changes (in a book like this, at least).

In summary, light from a single source generates an interference pattern when it passes through two slits. The *intensity* of the light varies. And an easy way to view how it varies in a plane is to use a screen.

Now we get to the really surprising bit. When this experiment is repeated with electrons, which we tend to think of as particles, *the same kind of pattern appears*. Even more surprising is that when electrons are fired *one at a time*, the pattern is *still* built up. (The positions of impact can be recorded.) So there are some places that an individual electron can go, and other places that it can't go. But quantum theory does not predict exactly where each individual electron will go. It only gives a *probability* that it will go to each place that it can go.

So what does this tell us about the world? Enter the interpretation of probability. Let's contrast thinking in terms of single case propensities with thinking in terms of information-based probabilities. If we use the former, we should conclude

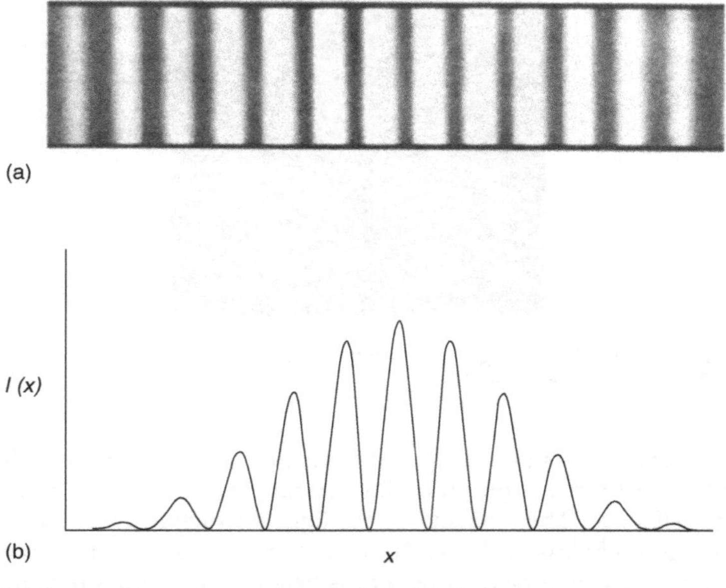

FIGURE 10.8 Image and intensity of light on a screen in Young's interference experiment

that *indeterminism* holds. There is no fact of the matter about which slit any individual electron will go through, or where it will appear on the screen, given its initial state (i.e. its state after leaving the electron gun). The outcome is not *uniquely determined* by, although it is *constrained by*, the initial conditions and the laws of nature. (It is 'constrained' in so far as some outcomes are *impossible*.)

What if we take the probabilities to be information-based, instead? Then we may continue to think that *determinism* holds. We may think that there *is* a fact of the matter about which slit any individual electron will go through, given its initial state. Moreover, we may think that there is a fact of the matter about the specific trajectory that it will follow. But we may think that this fact of the matter *is beyond our ability to determine*, because it is also *beyond our ability to determine the initial state of the electron with sufficient accuracy*. We may say that the systems set up in such electron interference experiments are *chaotic*, in so far as small (and unmeasurable) differences in initial conditions result in big differences in outcomes.

You may wonder which option is correct. However, this is a matter of great controversy (which I spent many of my early university years worrying about). My view now is that we just don't know. (So I don't think that quantum theory provides evidence that the world is indeterministic, even assuming it is true. I do, however, believe that it's nice to *think* in the deterministic way, when learning quantum theory. This makes the theory more accessible and comprehensible, in comparison with classical and folk physics, than thinking in terms of single case propensities does.) If you study physics, and are interested in learning more about this, then Cushing (1994) is a great place to start. Maybe you, too, will like Bohm's (deterministic) interpretation of quantum mechanics.

5 The Final Curtain

This final chapter has given an indication of how widely probabilities are used, in science and beyond. It has also shown that how we (can) interpret the probabilities in theories has a considerable bearing on how we (can) understand the theories. Sometimes, a theory can be predictively successful although it is unclear how the probabilities therein should be interpreted; quantum theory, for example, is like this. On other occasions, the success of a theory is explicable because the probabilities therein should be understood in one particular way; plausibly, this is the situation with Mendelian genetics. In still other cases, it seems like understanding the probability-talk in different ways makes the theory useable in different ways, i.e. improves its applicability; we saw this when we considered game theory.

Further Reading

For an intermediate-level introduction to confirmation theory, see Hacking (2001). One of the key intermediate-advanced texts, defending a subjective approach, is Howson and Urbach (2005).

For an intermediate-level introduction to game theory, see Tadelis (2013). For an interesting example of an advanced discussion of the interpretation of probability in game theory, see Kadane and Larkey (1982) and Harsanyi (1982).

For an introduction to Mendelian genetics, see Griffiths et al. (2000: ch. 2). (You can find an excerpt from this chapter, on Mendel's experiments, at: http://www.ncbi.nlm.nih.gov/books/NBK22098/.) For more on the interpretation of probability in biology, at an intermediate-to-advanced level, see Millstein (2003) on evolutionary theory.

For an introduction to quantum mechanics, see Albert (1994); Cushing (1994) gives the most accessible introduction to Bohm's version. Bach et al. (2013) present a recent experiment on electron diffraction in a rather accessible manner. There are numerous research papers on probability in quantum mechanics, which typically focus on one version of the theory or another, and it is difficult to recommend a few in particular. (Do a search for 'probability quantum mechanics' on www.philpapers.org to see what I mean.)

Appendices

A: The Axioms and Laws of Probability

Probability theory was developed in the seventeenth century, as explained in Chapter 2. In the centuries that followed, probabilities were widely used; and the key laws governing them, namely the laws of addition and multiplication, were well known. Probability theory was not rigorously axiomatized, however, until the twentieth century.

The first axiomatic system was proposed by Andrey Kolmogorov, a mathematician, in 1933. And this is the most famous. But there are many other alternatives. And I will use such an alternative here, because Kolmogorov's axioms do not include the law of multiplication, or any statements about conditional probabilities. (For a full explanation of conditional probabilities, consult Chapter 3.) Kolmogorov uses his axioms, which concern only unconditional probabilities, to *define* conditional probabilities. But this assumes that talk about conditional probabilities is *really* talk about unconditional probabilities. I prefer to treat conditional probabilities as fundamental, and primitive, in line with the information-based interpretations covered in Chapters 3 and 4 (where relationships between propositions or beliefs are central), as well as the world-based interpretations (where relationships between attributes and collectives or outcomes and repeatable conditions are central) covered in Chapters 7 and 8.

The axioms below are based on those preferred by De Finetti, and used by Gillies (2000: ch. 4). One key difference, however, is that they do not presume that probabilities concern events, rather than propositions.

1 $0 \leq P(E) \leq 1$, for any event (or proposition) E.
2 $P(\Omega) = 1$, when Ω is a certain event (or proposition).
3 $P(E_1) + \ldots + P(E_n) = 1$, when $E_1, \ldots, E_n$ are mutually exclusive and jointly exhaustive events (or propositions).
4 $P(E \,\&\, F) = P(E, F)P(F)$, when E and F are any two events (or propositions).

The phrases 'mutually exclusive' and 'jointly exhaustive', which appear in the third axiom, require explanation. On the one hand, two (or more) events are mutually exclusive if only one can occur. Similarly, two (or more) propositions are mutually exclusive if only one can be true. On the other hand, two (or more) events are jointly exhaustive if at least one must occur. And two (or more) propositions are jointly exhaustive if at least one must be true. Hence, an example of two mutually exclusive and jointly exhaustive events is 'England won the World Cup in 1966' and 'England did not win the World Cup in 1966'. An example of two mutually exclusive and jointly exhaustive propositions (which do not represent events) is 'One plus one is two' and 'One plus one is not two'.

Axiom 3 is *The Addition Law*. As shown by Gillies (2000: 59–60), it can also be stated as:

> $P(E \text{ or } F) = P(E) + P(F)$, where E and F are any two mutually exclusive events (or propositions).

The addition law also has a more general form, which applies even when E and F are not mutually exclusive:

> $P(E \text{ or } F) = P(E) + P(F) - P(E \,\&\, F)$, where E and F are any two events (or propositions).

Axiom 4 is *The Multiplication Law*. It has the following special form, when E and F are independent:

> $P(E \,\&\, F) = P(E)P(F)$, when E and F are any two independent events (or propositions).

E and *F* are independent events if whether *E* occurs has no bearing on whether *F* occurs, and vice versa. *E* and *F* are independent propositions if the truth of one has no bearing on the truth of the other, and vice versa. Examples of independent events are 'You put on red trousers tomorrow morning' and 'I eat pasta today for lunch'. (The results of coin flips are also typically taken to be independent.) Examples of independent propositions are 'Two multiplied by two is four' and 'Paris is the capital of France'.

Appendices
B: Bayes's Theorem

Bayes's theorem is a result of the axioms of probability. It has been used to develop accounts of how our beliefs should change, over time, and also, on a related note, how theories can be confirmed in science.

Imagine we are discussing some hypothesis, h, and some evidence, e. In everyday life, h could be 'It will rain in Hong Kong tomorrow' and e could be 'It rained in Hong Kong today'. In science, h could be relativity theory, and e could be 'Atomic clocks flown around the world on aircraft become out of step with clocks left on Earth during the same period.'

Bayes's theorem, as it pertains to any such h and e, can be written as follows:

$$P(h|e) = \frac{P(h)P(e|h)}{P(e)}$$

$P(h|e)$ is the *posterior probability* of h. It is the probability of h in the presence of e, i.e. assuming that e is true.

$P(h)$ is the *prior probability* of h. It is the probability of h in the absence of e, e.g. before e was discovered or even considered.

$P(e|h)$ is the *likelihood* of e on h. It is the probability of e on the assumption that h is true. It measures to what extent h predicts e.

Finally, $P(e)$ is the *marginal likelihood* of e. It is the probability of e without assuming h, i.e. independently of whether h is true.

To see how Bayes's theorem works, let's consider a simple scenario.

You are on a quiz show. You are shown the contents of two bags, A and B. A has three black rabbits in it. B has two black rabbits and one white rabbit in it. The bags are then sealed, so that you cannot see into either.

Next, the bags are shuffled and one is selected at random. The host of the show randomly draws out two rabbits from the bag, and sets each free to run around in the studio. They are both black! Then comes the question you have been waiting for: 'Is the last rabbit in the bag black? Yes or no?' If you answer it correctly, you will win a million US dollars! What should you do?

You know that the correct answer depends on which bag the host has randomly picked. If it's bag A, the answer is 'Yes'. If it's bag B, the answer is 'No'. But which bag did he pick? If you knew the value of $P(h|e)$, when h is 'The bag is A' and e is 'Two black rabbits were removed', it would help. But it is very difficult to see what that value is.

Enter Bayes's theorem, which gives a neat way of calculating $P(h|e)$. Let's work through the calculation.

(1) The initial pick between A and B was random. So the host was just as likely to get A as B. Thus $P(h) = ½$
(2) $P(h|e)$ is the probability of randomly picking two black rabbits *on the assumption* that the bag selected initially was A. And since only black rabbits are in that bag, we have to conclude that black rabbits will always be drawn from it. So $P(e|h) = 1$.

So far so good; we only need one more value to plug into the right hand side of Bayes's theorem, namely $P(e)$, in order to work out $P(h|e)$. But what is $P(e)$?

It will help us to note the following result from the axioms of probability:

$$P(e) = P(h)P(e|h) + P(\neg h)P(e|\neg h)$$

Hence, we can rewrite Bayes's theorem as:

$$P(h|e) = \frac{P(h)P(e|h)}{P(h)P(e|h) + P(\neg h)P(e|\neg h)}$$

Now we can proceed easily:

(3) $P(h)P(e|h)$ may be calculated from the results of steps (1) and (2).

$$P(h)P(e|h) = \tfrac{1}{2} * 1 = \tfrac{1}{2}$$

(4) If A was not the initial bag selected, then B must have been selected. So $\sim h$ is effectively the hypothesis that bag B was selected. And as noted in (1), the initial pick between A and B was random. So $P(\neg h) = \tfrac{1}{2}$

(5) $P(e|\neg h)$ is the probability of randomly picking two rabbits *on the assumption* that the bag selected initially was B. It is the probability of drawing a black rabbit on the first pick (when 2/3 of the rabbits are black) multiplied by the probability of drawing a black rabbit on the next pick (when 1/2 of the rabbits are black). Hence, $P(e|\neg h) = \frac{2}{3} * \frac{1}{2} = \frac{1}{3}$

We have everything we need.

$$P(h|e) = \frac{\frac{1}{2}}{\frac{1}{2} + \frac{1}{2} * \frac{1}{3}}$$

$$= \frac{\frac{1}{2}}{\frac{1}{2} + \frac{1}{6}}$$

$$= \frac{3}{4}$$

You are right to expect a black rabbit rather than a white rabbit! And you are one step closer to winning the million dollars...

References

Achinstein, P. 1995. 'Are Empirical Evidence Claims A Priori?', *British Journal for the Philosophy of Science* 46, 447–473.

Albert, D. Z. 1994. *Quantum Mechanics and Experience*. Harvard: Harvard University Press.

Bach, R., D. Pope, S.-H. Liou, and H. Batelaan. 2013. 'Controlled Double-Slit Electron Diffraction', *New Journal of Physics* 15, 033018. (doi:10.1088/1367-2630/15/3/033018)

Bertrand, J. 1888. *Calcul des Probabilités*. Paris: Gauthier-Villars.

Bridgman, P. 1927. *The Logic of Modern Physics*. New York: MacMillan.

Carnap, R. 1950. *Logical Foundations of Probability*. Chicago: University of Chicago Press.

Childers, T. 2013. *Philosophy and Probability*. Oxford: Oxford University Press.

Cushing, J. 1994. *Quantum Mechanics: Historical Contingency and the Copenhagen Hegemony*. Chicago: University of Chicago Press.

Daston, L. 1988. *Classical Probability in the Enlightenment*. Princeton: Princeton University Press.

David, F. N. 1962. *Games, Gods, and Gambling: The Origins and History of Probability and Statistical Ideas from the Earliest Times to the Newtonian Era*. New York: Hafner Press.

De Finetti, B. 1937. 'Foresight: Its Logical Laws, Its Subjective Sources', in H. E. Kyburg and H. E. Smokler (eds), *Studies in Subjective Probability*. New York: Wiley, pp. 93–158.

De Finetti, B. 1990. *Theory of Probability, Vol. I*. New York: Wiley.

Eagle, A. (ed.) 2011. *Philosophy of Probability: Contemporary Readings*. London: Routledge.

Eagle, A. 2004. 'Twenty-One Arguments against Propensity Analyses of Probability', *Erkenntnis* 60, 371–416.

Eriksson, L. and A. Hájek. 2007. 'What are Degrees of Belief?', *Studia Logica* 86, 183–213.

Fetzer, J. H. 1981. *Scientific Knowledge: Causation, Explanation, and Corroboration*. Dordrecht: D. Reidel.

Fetzer, J. H. 1982. 'Probabilistic Explanations', *PSA: Proceedings of the Biennial Meeting of the Philosophy of Science Association* 1982, 194–207.

Fetzer, J. H. 1988. 'Probabilistic Metaphysics', in J. H. Fetzer (ed.), *Probability and Causality*. Dordrecht: D. Reidel, pp. 109–132.

Fiedler, K. 1988. 'The Dependence of the Conjunction Fallacy on Subtle Linguistic Factors', *Psychological Research* 50, 123–129.

Gillies, D. 1991. 'Intersubjective Probability and Confirmation Theory', *British Journal for the Philosophy of Science* 42, 513–533.

Gillies, D. 2000. *Philosophical Theories of Probability*. London: Routledge.

Griffiths, A. J. F., J. H. Miller, D. T. Suzuki, R. C. Lewontin, and W. M. Gelbart. 2000. *An Introduction to Genetic Analysis*. New York: W. H. Freeman.

Hacking, I. 1975. *The Emergence of Probability: A Philosophical Study of Early Ideas about Probability, Induction and Statistical Inference*. Cambridge: Cambridge University Press.

Hacking, I. 1987. 'The Inverse Gambler's Fallacy: The Argument from Design. The Anthropic Principle Applied to Wheeler Universes', *Mind* 96, 331–340.

Hacking, I. 2001. *An Introduction to Probability and Inductive Logic*. Cambridge: Cambridge University Press.

Hájek, A. 1997. ' "Mises Redux" – Redux: Fifteen Arguments Against Finite Frequentism', *Erkenntnis* 45, 209–227.

Hájek, A. 2009. 'Fifteen Arguments Against Hypothetical Frequentism', *Erkenntnis* 70, 211–235.

Handfield, T. 2012. *A Philosophical Guide to Chance: Physical Probability*. Cambridge: Cambridge University Press.

Harsanyi, J. C. 1982. 'Subjective Probability and the Theory of Games: Comments on Kadane and Larkey's Paper', *Management Science* 28, 120–124.

Howson, C. and P. Urbach. 2005. *Scientific Reasoning: The Bayesian Approach*. La Salle: Open Court.

Humphreys, P. 1985. 'Why Propensities Cannot Be Probabilities', *The Philosophical Review* 94, 557–570.

Humphreys, P. 1989. *The Chances of Explanation: Causal Explanation in the Social, Medical and Physical Sciences*. Princeton: Princeton University Press.

Jaynes, E. T. 1957. 'Information Theory and Statistical Mechanics', *Physical Review* 106, 620–630.

Jaynes, E. T. 2003. *Probability Theory: The Logic of Science*. Cambridge: Cambridge University Press.

Jeffrey, R. 2004. *Subjective Probability: The Real Thing*. Cambridge: Cambridge University Press.

Kadane, J. B. and P. D. Larkey. 1982. 'Subjective Probability and the Theory of Games', *Management Science* 28, 113–120.

Kalinowski, P., F. Fidler, and G. Cumming. 2008. 'Overcoming the Inverse Probability Fallacy: A Comparison of Two Teaching Interventions', *Methodology* 4, 152–158.

Keynes, J. M. 1921. *A Treatise on Probability*. London: Macmillan.

Koehler, J. 1996. 'The Base Rate Fallacy Reconsidered: Descriptive, Normative, and Methodological Challenges', *Behavioral and Brain Sciences* 19, 1–53.

Kyburg, H. E. 1970. *Probability and Inductive Logic*. London: Macmillan.

Laplace, P.-S. 1814 (English edition 1951). *A Philosophical Essay on Probabilities*. New York: Dover Publications Inc.

Mikkelson, J. 2004. 'Dissolving the Wine/Water Paradox', *British Journal for the Philosophy of Science* 55, 137–145.

Miller, D. W. 1994. *Critical Rationalism: A Restatement and Defence*. La Salle: Open Court.

Millstein, R. L. 2003. 'Interpretations of Probability in Evolutionary Theory', *Philosophy of Science* 70, 1317–1328.

Popper, K. R. 1957. 'The Propensity Interpretation of the Calculus of Probability, and the Quantum Theory', in S. Körner (ed.), *Observation and Interpretation: A Symposium of Philosophers and Physicists*. London: Butterworths, pp. 65–70 and 88–89.

Popper, K. R. 1959a. 'The Propensity Interpretation of Probability', *British Journal for the Philosophy of Science* 10, 25–42.

Popper, K. R. 1959b. *The Logic of Scientific Discovery*. New York: Basic Books.

Popper, K. R. 1967. 'Quantum Mechanics without "The Observer"', in M. Bunge (ed.), *Quantum Theory and Reality*. New York: Springer, pp. 7–44.

Popper, K. R. 1983. *Realism and the Aim of Science*. London: Routledge.

Popper, K. R. 1990. *A World of Propensities*. Bristol: Thoemmes.

Ramsey, F. P. 1926. 'Truth and Probability', in F. P. Ramsey, *The Foundations of Mathematics and other Logical Essays*, ed. R. B. Braithwaite. London: Kegan Paul, Trench, Trübner & Co., 1931, pp.156–198.

Reinhart, A. 2015. *Statistics Done Wrong: The Woefully Complete Guide*. San Francisco, CA: No Starch Press.

Rowbottom, D. P. 2008. 'On the Proximity of the Logical and "Objective Bayesian" Interpretations of Probability', *Erkenntnis* 69, 335–349.

Rowbottom, D. P. 2013a. 'Empirical Evidence Claims Are A Priori', *Synthese* 19, 2821–2834.

Rowbottom, D. P. 2013b. 'Group Level Interpretations of Probability: New Directions', *Pacific Philosophical Quarterly* 94, 188–203.

Selvin, S. 1975. 'On the Monty Hall Problem', *American Statistician* 29, 134.

Suárez, M. 2013. 'Propensities and Pragmatism', *Journal of Philosophy* CX, 61–92.

Tadelis, S. 2013. *Game Theory: An Introduction*. Princeton: Princeton University Press.

Tversky, A. and D. Kahneman. 1982. 'Judgments of and by Representativeness', in D. Kahneman, P. Slovic, and A. Tversky (eds), *Judgment Under Uncertainty: Heuristics and Biases*. Cambridge: Cambridge University Press, pp. 84–98.

Tversky, A. and D. Kahneman. 1983. 'Extensional Versus Intuitive Reasoning: The Conjunction Fallacy in Probability Judgment', *Psychological Review* 90, 293–315.

Von Mises, R. 1928/1968. *Probability, Statistics and Truth*. 2nd ed. London: Allen and Unwin.

Vos Savant, M. 2014. 'Game Show Problem', http://marilynvossavant.com/game-show-problem/

Williamson, J. 2010. *In Defence of Objective Bayesianism*. Oxford: Oxford University Press.

Index

N.B. Instances of names appearing in the Preface, Further Reading sections and References section are not indexed.